Tools for High Performance Computing 2015

Andreas Knüpfer · Tobias Hilbrich
Christoph Niethammer · José Gracia
Wolfgang E. Nagel · Michael M. Resch
Editors

Tools for High Performance Computing 2015

Proceedings of the 9th International Workshop on Parallel Tools for High Performance Computing, September 2015, Dresden, Germany

Springer

Editors
Andreas Knüpfer
Zentrum für Informationsdienste und Hochleistungsrechnen (ZIH)
Technische Universität Dresden
Dresden
Germany

Tobias Hilbrich
Zentrum für Informationsdienste und Hochleistungsrechnen (ZIH)
Technische Universität Dresden
Dresden
Germany

Christoph Niethammer
Höchstleistungszentrum Stuttgart (HLRS)
Universität Stuttgart
Stuttgart
Germany

José Gracia
Höchstleistungszentrum Stuttgart (HLRS)
Universität Stuttgart
Stuttgart
Germany

Wolfgang E. Nagel
Zentrum für Informationsdienste und Hochleistungsrechnen (ZIH)
Technische Universität Dresden
Dresden
Germany

Michael M. Resch
Höchstleistungszentrum Stuttgart (HLRS)
Universität Stuttgart
Stuttgart
Germany

Cover front figure: OpenFOAM Large Eddy Simulations of dimethyl ether combustion with growing resolutions of 1.3 million elements, 10 million elements, and 100 million elements from left to right reveal how more computing power produces more realistic results. Courtesy of Sebastian Popp, Prof. Christian Hasse, TU Bergakademie Freiberg, Germany.

ISBN 978-3-319-39588-3   ISBN 978-3-319-39589-0 (eBook)
DOI 10.1007/978-3-319-39589-0

Library of Congress Control Number: 2016941316

Mathematics Subject Classification (2010): 68U20

Printed on acid-free paper

This Springer imprint is published by Springer Nature
The registered company is Springer International Publishing AG Switzerland

Preface

Highest-scale parallel computing remains a challenging task that offers huge potentials and benefits for science and society. At the same time, it requires deep understanding of the computational matters and specialized software in order to use it effectively and efficiently.

Maybe the most prominent challenge nowadays, on the hardware side, is heterogeneity in High Performance Computing (HPC) architectures. This inflicts challenges on the software side. First, it adds complexity for parallel programming, because one parallelization model is not enough; rather two or three need to be combined. And second, portability and especially performance portability are at risk. Developers need to decide which architectures they want to support. Development or effort decisions can exclude certain architectures. Also, developers need to consider specific performance tuning for their target hardware architecture, which may cause performance penalties on others. Yet, avoiding architecture specific optimizations altogether is also a performance loss, compared to a single specific optimization. As the last resort, one can maintain a set of specific variants of the same code. This is unsatisfactory in terms of software development and it multiplies the necessary effort for testing, debugging, performance analysis, tuning, etc. Other challenges in HPC remain relevant such as reliability, energy efficiency, or reproducibility.

Dedicated software tools are still important parts of the HPC software landscape to relieve or solve today's challenges. Even though a *tool* is by definition not a part of an application, but rather a supplemental piece of software, it can make a fundamental difference during the development of an application. This starts with a debugger that makes it possible (or just more convenient and quicker) to detect a critical mistake. And it goes all the way to performance analysis tools that help to speed up or scale up the application, potentially resolving system effects that could not be understood without the tool. Software tools in HPC face their own challenges. In addition to the general challenges mentioned above there is the *bootstrap challenge*—tools should be there early when a new hardware architecture is introduced or an unprecedented scalability level is reached. Yet, there are no tools to help the tools to get there.

Since the previous workshop in this series, there have been interesting developments for stable and reliable tools as well as tool frameworks. Also there are new approaches and experimental tools that are still under research. Both kinds are very valuable for a software ecosystem, of course. In addition, there are greatly appreciated verification activities for existing tools components. And there are valuable standardization efforts for tools interfaces in parallel programming abstractions.

The 9th International Parallel Tools Workshop in Dresden in September 2015 included all those topics. In addition, there was a special session about user experiences with tools including a panel discussion. And as an outreach to another community of computation intensive science there was a session about Big Data algorithms. The contributions presented there are interesting in two ways. First as target applications for HPC tools. And second as interesting methods that may be employed in the HPC tools.

This book contains the contributed papers to the presentations at the workshop in September 2015.[1] As in the previous years, the workshop was organized jointly between the Center of Information Services and High Performance Computing (ZIH)[2] and the High Performance Computing Center (HLRS).[3]

Dresden, Germany
January 2016

Andreas Knüpfer
Tobias Hilbrich
Christoph Niethammer
José Gracia
Wolfgang E. Nagel
Michael M. Resch

[1] http://tools.zih.tu-dresden.de/2015/.

[2] http://tu-dresden.de/zih/.

[3] http://www.hlrs.de.

Contents

Chapter 1
Dyninst and MRNet: Foundational Infrastructure for Parallel Tools

William R. Williams, Xiaozhu Meng, Benjamin Welton and Barton P. Miller

Abstract Parallel tools require common pieces of infrastructure: the ability to control, monitor, and instrument programs, and the ability to massively scale these operations as the application program being studied scales. The Paradyn Project has a long history of developing new technologies in these two areas and producing ready-to-use tool kits that embody these technologies: Dyninst, which provides binary program control, instrumentation, and modification, and MRNet, which provides a scalable and extensible infrastructure to simplify the construction of massively parallel tools, middleware and applications. We will discuss new techniques that we have developed in these areas, and present examples of current use of these tool kits in a variety of tool and middleware projects. In addition, we will discuss features in these tool kits that have not yet been fully exploited in parallel tool development, and that could lead to advancements in parallel tools.

1.1 Introduction

Parallel tools require common pieces of infrastructure: the ability to control, monitor, and instrument programs, and the ability to massively scale these operations as the application program being studied scales. The Paradyn Project has a long history of developing new technologies in these two areas and producing ready-to-use tool kits that embody these technologies. One of these tool kits is Dyninst, which provides binary program control, instrumentation, and modification. When we initially designed Dyninst, our goal was to provide a platform-independent binary instrumentation platform that captured only the necessary complexities of binary code. We believe that the breadth of tools using Dyninst, and the breadth of Dyninst components that they use, reflects how well we have adhered to these guiding principles. We discuss the structure and features of Dyninst in Sect. 1.2.

Another tool kit we have developed is MRNet, which provides a scalable and extensible infrastructure to simplify the construction of massively parallel tools,

W.R. Williams (✉) · X. Meng · B. Welton · B.P. Miller
University of Wisconsin, 1210 W. Dayton St., Madison, WI 53706, USA
e-mail: bill@cs.wisc.edu

A. Knüpfer et al. (eds.), *Tools for High Performance Computing 2015*,
DOI 10.1007/978-3-319-39589-0_1

middleware and applications. MRNet was designed from the beginning to be a flexible and scalable piece of infrastructure for a wide variety of tools. It has been applied to data aggregation, command and control, and even to the implementation of distributed filesystems. MRNet provides the scalability foundation for several critical pieces of debugging software. We discuss the features of MRNet in Sect. 1.3.

We discuss common problems in scalable tool development that our tool kits have been used to solve in the domains of performance analysis (Sect. 1.4) and debugging (Sect. 1.5). These problems include providing control flow context for an address in the binary, providing local variable locations and values that are valid at an address in the binary, collecting execution and stack traces, aggregating trace data, and dynamically instrumenting a binary in response to newly collected information.

We also discuss several usage scenarios of our tool kits in binary analysis (Sect. 1.6) and binary modification (Sect. 1.7) applications. Analysis applications of our tools (Fig. 1.3) include enhancing debugging information to provide a more accurate mapping of memory and register locations to local variables, improved analysis of indirect branches, and improved detection of function entry points that lack symbol information. Applications of our tools for binary modification include instruction replacement, control flow graph modification, and stack layout modification. Some of these analysis and modification applications have already proven useful in high-performance computing. We conclude (Sect. 1.8) with a summary of future plans for development.

1.2 DyninstAPI and Components

DyninstAPI provides an interface for binary instrumentation, modification, and control, operating both on running processes and on binary files (executables and libraries) on disk. Its fundamental abstractions are *points*, specifying where to instrument, and *snippets*, specifying what instrumentation should do. Dyninst provides platform-independent abstractions representing many aspects of processes and binaries, including address spaces, functions, variables, basic blocks, control flow edges, binary files and their component modules.

Points are specified in terms of the control flow graph (CFG) of a binary. This provides a natural description of locations that programmers understand, such as function entry/exit, loop entry/exit, basic block boundaries, call sites, and control flow edges. Previous work, including earlier versions of Dyninst [7], specified instrumentation locations by instruction addresses or by control flow transfers. Bernat and Miller [5] provide a detailed argument why, in general, instrumentation before or after an instruction, or instrumentation on a control transfer, does not accurately capture certain important locations in the program. In particular, it is difficult to characterize points related to functions or loops by using only addresses or control transfers.

Snippets are specified in a platform-independent abstract syntax tree language [7]. The platform-independent nature of the instrumentation specification allows Dyninst-based tools (*mutators*) to, in most cases, be written once and run on any supported platform.

To instrument a binary, extra space must be provided in the code for the instrumentation code. This space may be created by relocating some or all of the original code in order to provide room for instrumentation. The instrumentation and associated program code may be positioned so that the instrumentation executes inline with its program context or out-of-line from its context. Bernat and Miller [5] determined that, given current processor characteristics, relocating whole functions and generating their associated instrumentation inline minimizes overhead by improving instruction cache coherence compared to other approaches.

Dyninst has been deconstructed into several component libraries [24], each performing some aspect of binary instrumentation, analysis, or control (Fig. 1.1). As we will see, many of these components are commonly used in various smaller subsets for common tasks in parallel tool design. As a benefit of creating smaller components, each of these components deals with a much smaller amount of platform variation than Dyninst. For example, while Dyninst supports a wide variety of architectures and operating systems, the SymtabAPI component is concerned primarily with the details of binary file formats. This allows us to largely simplify SymtabAPI to handling ELF and PE files correctly, with small and well-defined architecture and

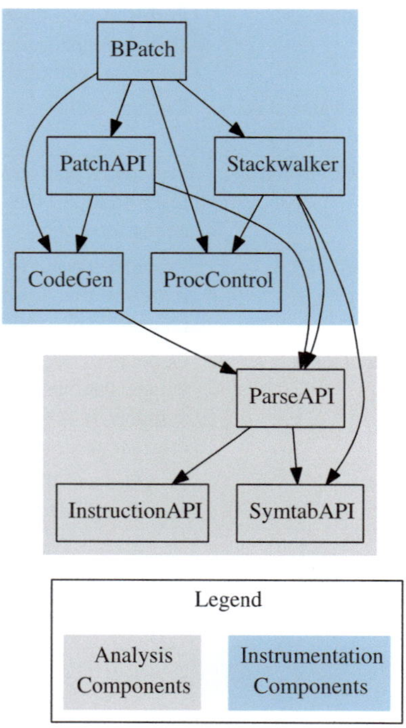

Fig. 1.1 Dyninst and its component libraries

operating system specific subcomponents. The Dyninst components include tools for analyzing and interpreting binaries, interacting with processes, and modifying binaries and inserting instrumentation. The analysis and interpretation components include SymtabAPI, which provides a format-independent representation of binary files and debugging information; InstructionAPI, which disassembles instructions; ParseAPI, which constructs control flow graphs; and DataflowAPI, which contains a selection of data flow analysis algorithms used inside Dyninst. StackwalkerAPI and ProcControlAPI, respectively, collect stack traces from processes and control processes and threads via the debug interface of the operating system. PatchAPI, CodeGen, DynC, and DyninstAPI itself collectively provide the point-snippet interface used by instrumentation, the interfaces for control flow modification, and a C-like wrapper language to generate snippet construction code. The components and their supported platforms are listed in Table 1.1.

Table 1.1 Dyninst components and their capabilities

Component	Description	Supported platforms
SymtabAPI	Reads symbol tables and debugging information	ELF, PE
InstructionAPI	Decodes instructions to an operation and operand ASTs	x86, x86_64, PowerPC32, PowerPC64
ParseAPI	Constructs control flow graphs	x86, x86_64, PowerPC32, PowerPC64
DataflowAPI	Performs data flow analyses: slicing, register liveness, stack analysis, symbolic evaluation	x86, x86_64, PowerPC32, PowerPC64
StackwalkerAPI	Collects call stacks	Linux, Windows, x86, x86_64, PowerPC32, PowerPC64, ARMv8
ProcControlAPI	Provides a platform-independent layer on top of the operating system debug interface	Linux, Windows, x86, x86_64, PowerPC32, PowerPC64, ARMv8
PatchAPI	Provides a point of indirection to represent transformations to a control flow graph	x86, x86_64, PowerPC32, PowerPC64
CodeGen	Generates code for instrumentation snippets and code to ensure those snippets do not interfere with the original program	x86, x86_64, PowerPC32, PowerPC64
DynC	Provides a C-like language for specifying instrumentation snippets	x86, x86_64, PowerPC32, PowerPC64

1.3 MRNet

Scalable computation is an important challenge, whether you are building applications, tools, or large scale distributed systems. The challenge of scale requires that developers for distributed systems select computational patterns that have the properties that allow for scaling and the expressiveness to apply to a broad range of problems. Tree-based Overlay Networks (TBONs) are an ideal method of parallelizing computation supplying a scalable communication pattern that can express the solution to a wide range of distributed computation problems. TBONs connect a set of processes into a tree layout where leaf nodes perform the bulk processing work, internal tree processes perform aggregation/multicasting of results out of the tree, and a single front end process which aggregates results to produce a single output. Scalability is achieved with TBONs by use of aggregation and multicast filters to reduce data moving through the tree.

The Multicast Reduction Network (MRNet) [26] is a framework that implements the TBON model to provide scalable communication to distributed system developers. MRNet handles the creation and connection of processes into a tree network layout. MRNet assigns each process a role as a *frontend* (FE), *communication* (CP), or *backend* (BE) process, as shown in Fig. 1.2. The size of the tree and layout of

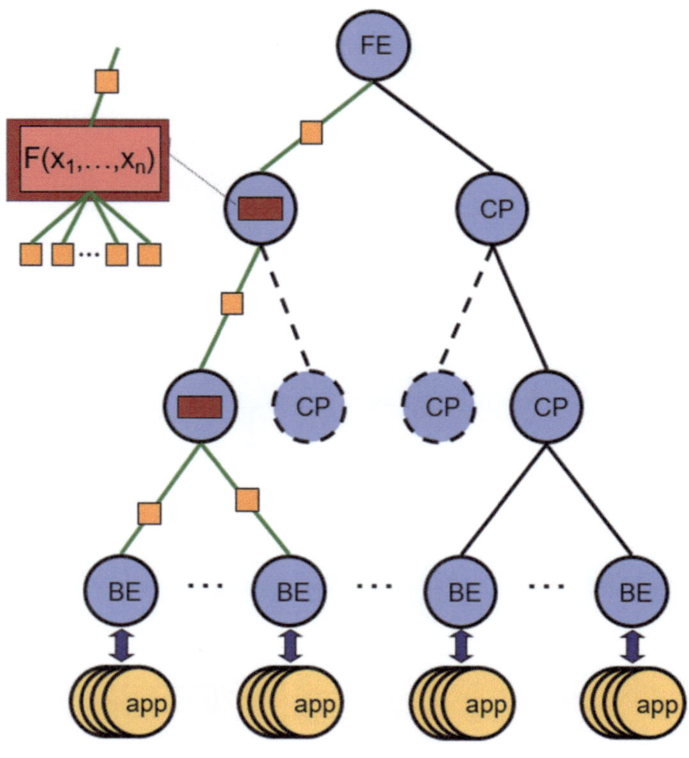

Fig. 1.2 The layout of a MRNet tree and its various components

processes can be modified by users without modifying the program, allowing a single codebase to scale from one process to millions. Users can supply custom aggregation and multicast filters to MRNet. The MRNet framework has been used extensively to build highly scalable tools and applications that are in use on leadership class machines [2, 3, 28].

1.4 Performance Tools

Performance tools collect and interpret information about how a program uses various system resources, such as CPU, memory, and networks. There are two notable categories of performance tools where Dyninst components have been used as part of these tasks: sampling tools and tracing tools. Figures 1.3 and 1.4 illustrate how performance tools may use Dyninst and its components in both analysis and instrumentation contexts.

1.4.1 Sampling Tools

Sampling tools periodically observe some aspect of program behavior and record these observations. One common form of sampling is call-stack sampling, which collects a set of program counter (PC) values and return address (RA) values that comprise the call stack of an executing thread. From these addresses in the program's code segment, one may derive a variety of further context:

- Binary file
- Source file
- Function

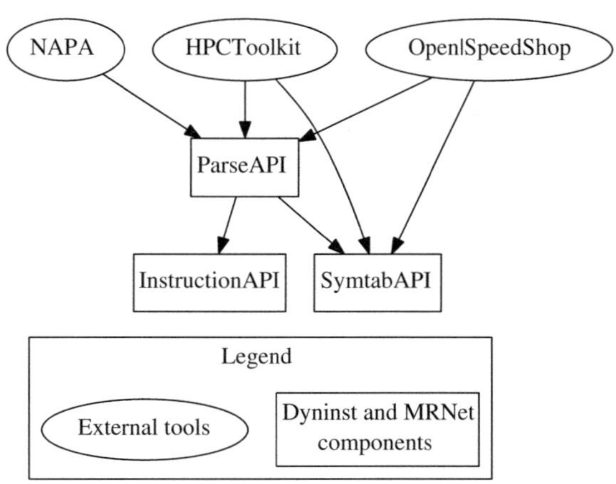

Fig. 1.3 Tools using Dyninst's binary analysis components

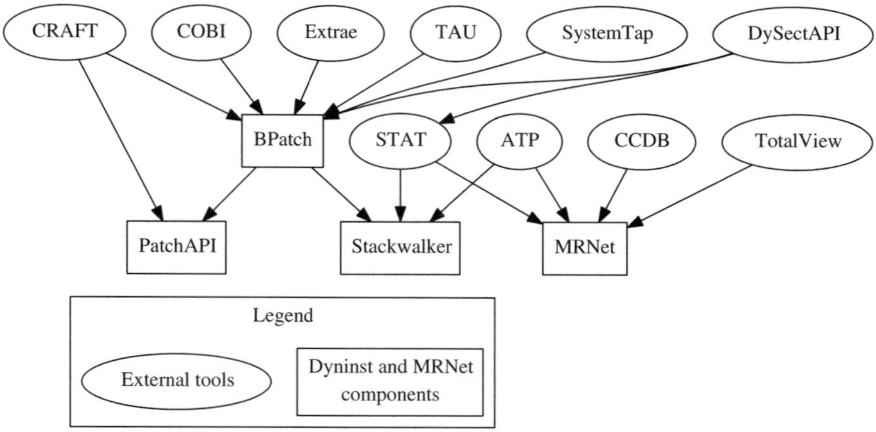

Fig. 1.4 Tools using Dyninst's instrumentation, monitoring, and debugging components

- Source line
- Loop
- Basic block

HPCToolkit [1] and Open|SpeedShop [29] both use SymtabAPI and ParseAPI to determine this contextual information from the addresses in a call stack. Reconstructing both the full source-level calling context (including inline functions) and the loop nesting context (including irreducible loops) from a call stack provides users with additional insight into where their code suffers from performance problems.

1.4.2 Tracing Tools

Tracing may be performed at function, basic block, memory reference, or instruction granularities. It captures records of events as they occur in the program. Many well-known performance tools collect or analyze tracing data. In particular, COBI [23], Tau [31], and Extrae [20] can use Dyninst's binary rewriting functionality in order to insert instrumentation that produces tracing data.

Instrumentation-based tracing relies on the insertion of instrumentation at the various points where trace data is to be collected. This instrumentation may be inserted as source code, during the compilation and linking process, through modification of the binary once it has been linked, or at run time. Instrumentation that occurs at any point up to and including the linking process we describe as source instrumentation; instrumentation that occurs afterward we describe as binary instrumentation. Dyninst and its components are concerned with binary instrumentation.

Binary instrumentation relies on the ability to understand and manipulate binary code without access to authoritative source code or compiler intermediate representations. It is necessarily a more difficult process than source instrumentation, but comes with compensating advantages. First, if a user wishes to dynamically modify the set of events that they are recording based on the data they have collected, it is necessary to insert or remove instrumentation at run time. In the subset of cases where the user already knows what instrumentation they would wish to insert prior to run time, this capability can be emulated at the source level by selectively enabling and disabling instrumentation execution, but binary instrumentation allows arbitrary instrumentation to be inserted at arbitrary points in response to observed events. Second, binary instrumentation can be inserted without recompiling a program. Particularly in cases where a user wishes to instrument a small portion of a large application, this can provide a significant time savings. Third, while source instrumenters have been improving their ability to avoid perturbing the resulting binary, they are not perfect in this respect. Binary instrumentation, of course, also perturbs the resulting binary, but its starting ground truth is precisely the binary that the compiler would generate without instrumentation.

In addition to tracing control flow events, the Dyninst interface allows users to perform tracing of a wide variety of memory operations: tracking allocations and deallocations, instrumenting memory accesses, and observing the effective addresses and byte counts they affect. As with all forms of fine-grained (instruction level) instrumentation, the overhead imposed by observing and recording every memory access is quite high in most cases. It is consequently common in our experience for users to develop specialized tools for memory tracing to diagnose particular performance problems. We hope that broader exposure of the Dyninst memory instrumentation features will lead to more general-purpose memory instrumentation tools being developed, both for performance analysis and for debugging.

1.5 Debugging Tools

In addition to analyzing the performance of parallel software, it is often necessary to debug it. The combination of certain Dyninst components with MRNet has proven to be a potent combination for developing lightweight tools for identifying bugs at extremely large scales.

1.5.1 Stack Trace Aggregation

A basic and useful approach to developing highly scalable debugging tools is stack trace aggregation: collecting stack traces from all of the threads and processes in a large parallel program, and merging them into a call stack prefix tree. Examples of this approach include Stack Trace Analysis Tool (STAT) [3] from Lawrence Livermore National Laboratories (LLNL) and Cray's Abnormal Termination Processing (ATP) tool [9]. Each of these tools uses StackwalkerAPI to collect call stacks. Users of the tools may collect local variable information and function contexts, as in Sect. 1.4.1, using SymtabAPI and potentially also ParseAPI. MRNet is then used by these tools to aggregate the stack traces in a scalable manner into a call stack prefix tree. STAT and ATP differ in their intended use cases; STAT is often used to debug hangs and stalls, whereas ATP is specifically focused on debugging crashes.

STAT has been successfully used to detect a wide variety of problems in both software and hardware. It has detected bugs in the LUSTRE filesystem, slow decrementors on particular processor cores resulting in 1,000,000x slowdowns in `sleep()`, and numerous bugs in application code as well [17]. STAT has collected call stacks from the entire Sequoia supercomputer (approximately 750,000 cores), and has collected call stacks from approximately 200k cores in under a second [19]. ATP is a standard part of Cray's Linux distribution [9], and is automatically invoked whenever an appropriately launched application crashes.

1.5.2 Distributed Debuggers with MRNet

MRNet has also been used as infrastructure for providing scalable control of existing full-featured debugging tools. The TotalView debugger has employed MRNet as a distributed process control layer [22], as has Cray's CCDB debugger.

TotalView is a high performance parallel debugger developed by Roguewave capable of debugging and profiling applications running on large node counts. With Totalview, application developers can perform wide range of debugging and profiling tasks such as setting breakpoints, reading and writing memory locations and registers, and single stepping through an application. MRNet is used by Totalview to scale these operations across an application running on thousands of nodes. A tree based overlay network is constructed between the application processes running on nodes and a frontend process that controls debugging and profiling operations. The frontend presents a user with a graphical representation of the current state of a running distributed application. A user can then issue commands (such as setting a breakpoint) that are passed through the overlay network down to application processes where they are executed. TotalView uses aggregation filters to reduce the volume of data generated by application processes so that a snapshot of the current state of a running application can be presented to the developer. Multicast filters are used by TotalView to broadcast commands down to individual nodes.

The Scalable Parallel Debugging Library [16] (SPDL), which provides a generic parallel debugging interface on top of MRNet and Eclipse SCI [8], has been used to extend Cray's CCDB debugger to larger scales [10]. SPDL provides comparable infrastructure to the TotalView implementation described above. CCDB, using this infrastructure, demonstrates command latency of less than a second at scales up to 32,000 processes.

1.5.3 Dynamic Instrumentation for Debugging

For some debugging problems, stack traces are insufficient, and the programmer requires knowledge of how the current point of execution was reached. This is an area where dynamic instrumentation can be applied in at least two ways: as a method for generating automated equivalents of typical interactive debugging commands, and as a method for generating debugging traces that precisely capture interesting behavior. We consider an example of each of these applications.

DySectAPI [15] builds on the foundation of STAT, and attempts to provide the ability to script `gdb`-like query and process control operations: breakpoints, probe points, conditional breakpoints and watchpoints, and access to registers and variables. Much of this functionality can be exposed with only trivial extensions to STAT (for instance, allowing the user to write to local variables as well as reading them); some, however, requires significantly more of the Dyninst component stack. In particular, the execution of an arbitrary remote procedure call requires some form of code generation.

SystemTap [12] is a kernel instrumentation and tracing tool developed by RedHat that uses Dyninst instrumentation to extend its capabilities to user space. The current SystemTap model is mostly oriented towards instrumentation specified statically, as it must support the compilation of scripts to kernel modules. For those cases where it is performing instrumentation that appears to be dynamic, that appearance is in most cases granted through conditional execution. SystemTap does allow scripts to invoke arbitrary system commands; we believe that special handling of the recursive invocation of SystemTap itself through dynamic instrumentation would increase the power of this idiom.

1.6 Analysis Tools

Improving the understanding of a binary's behavior can allow other tools to perform their tasks better. We present a data flow analysis use case, where slicing is used to improve the understanding of local variable access, and a control flow analysis use case, where accurate understanding of the CFG of a binary allows more efficient and accurate instrumentation within Dyninst itself.

1.6.1 Slicing

Slicing [33] is a data flow analysis that determines which instructions affect (backwards slicing) or are affected by (forwards slicing) the value of a given abstract location (register or memory location) at a given instruction. The DataflowAPI includes a slicing implementation that refines this concept to consider not just instructions, but assignments within those instructions.

The NAPA tool, currently under development at LLNL, uses DataflowAPI's slicer in an effort to improve the ability of tools to match individual load and store instructions with their corresponding variables. In principle, debugging information such as DWARF [11] should contain sufficient information that all such memory accesses can be resolved. In practice, for many data structures, this is not the case. For example, while the debugging information may contain one of the ways to refer to a location within an aggregate, the actual load or store will use a different alias to the same location. Applying a backwards slicing analysis to the load or store, searching through the containing function until the effective address being accessed has been derived from some set of local variables, improves the input data to further analyses, such as blame assignment [27].

1.6.2 Binary Parsing

The goal of parsing a binary is to represent the binary with code constructs that are familiar to programmers, including CFGs, functions, loops and basic blocks. These code constructs are the foundations for performing a data flow analysis, such as slicing (Sect. 1.6.1), and specifying instrumentation points, such as instrumenting at the entry of a function or at the exit of a loop.

Algorithms to recover these code constructs from binaries are encapsulated in ParseAPI. ParseAPI uses recursive traversal parsing [30] to construct basic blocks, determine function boundaries, and build CFGs. It starts from known entry points such as the program entry point and function entry points from symbol tables and follows the control flow transfers to build the CFG and identify more entry points. Not all code will necessarily be found by recursive traversal alone; this leaves *gaps* [14] in the binary where code may be present, but has not yet been identified. Furthermore, recursive traversal does not explicitly address the problem of how to resolve control flow targets in non-trivial cases, such as indirect branches. If these challenges are not handled properly, the parser would miss real code, have inaccurate CFGs, and observe degrading qualities of data flow analysis, binary instrumentation, and binary modification. We describe our new techniques for resolving jump tables, which represent a well-defined subset of indirect branches, and for gap parsing, which improves our parsing coverage for stripped binaries.

Jump tables are commonly used to implement switch statements and loop unrolling optimizations and they often represent intraprocedural control transfers. Because of Dyninst's function-based relocation approach (Sect. 1.2), it is necessary to safely overapproximate the potential targets of an indirect branch to relocate a function. This means that we must ensure that our understanding of a function's structure does not miss any code, and our understanding of its basic blocks does not ignore any block boundaries. In practical terms, this means that our analysis of an indirect branch must contain a proper superset of the true targets of that branch, or we will be unable to safely relocate and instrument the function containing the indirect branch.

We implemented a new slicing-based data flow analysis [21] to improve our handling of jump tables, relying on the following two key characterizations of jump tables: (1) jump table entries are contiguous and reside in read-only memory regions; (2) the jump target depends on a single bounded input value, which often corresponds to the switch variable in a switch statement. Our analysis is able to handle several variations of jump tables that appear in real software: (1) the table contents can be either jump target addresses or offsets relative to a base address; (2) the table location can be either explicitly encoded in instructions or computed; (3) the input value can be bounded through conditional jumps or computation; (4) arbitrary levels of tables involved in address calculation, where prior level tables are used to index into later level tables.

Our evaluations show that the new analysis can reduce the number of uninstrumentable functions in glibc by 30 % with a 20 % increase in parse overhead and reduce 7 % uninstrumentable functions in normal binaries with a 5 % increase in parse overhead.

Stripped binaries are significantly more difficult to analyze because when no function entry points are present, it is not easy to decide which addresses to start the control flow traversal. Recent research has used machine learning based approaches to learn code features such as instruction sequences [4, 25] or raw byte sequences [32] for identifying function entry points. Dyninst 9.0 uses Rosenblum et al's approach [25] to select instruction sequences from a set of training binaries and assigns each selected instruction sequence a weight to represent the probability that an address is a function entry point if the sequence is matched at the address. We scan through the binary searching for addresses where the probability that the address is a function entry point is greater than a configurable threshold. For each address where this is true, we then apply Dyninst's recursive traversal implementation, analyzing the function implied by this entry point and all of its callees to reduce the size of the gaps that must be scanned. Note that if we have identified a function entry point with some probability p, every one of its call targets must be a function entry point with probability $q \geq p$. Thus, all of the function entry points generated by this approach will be true function entry points with $p \geq t$ for a threshold t.

We compared the abilities of two versions of Dyninst to identify function entry points in stripped binaries. Dyninst 8.2.1 uses a few manually-designed instruction patterns and Dyninst 9.0 uses the machine learning approach to train its model. The test binaries are from binutils, coreutils, and findutils, built with ICC and GCC, at -O0 to -O3. The test results are summarized in Table 1.2. *Precision*, in this case, is the percentage of function entry points identified by Dyninst that are real function entry points; *recall* is the percentage of real function entry points identified as such.

We make two observations about these results. First, we see that the machine learning approach dramatically increases the recall in both 32-bit and 64-bit binaries, at the cost of some precision. This means that ParseAPI can discover much more code in gaps, with some of the discovered code being not real code. Second, the results

Table 1.2 Gap parsing test results

Version	Platform	Average precision (%)	Average recall (%)
Manually-designed patterns (8.2.1)	64-bit x86	98.1	37.4
Manually-designed patterns (8.2.1)	32-bit x86	95.6	53.9
Machine learning-derived idioms (9.0)	64-bit x86	94.7	83.2
Machine learning-derived idioms (9.0)	32-bit x86	97.1	93.8

show that 64-bit function entry points are more difficult to identify. Our examination of the rules generated for Dyninst 9.0 suggests that the increased size of the register set and the consequent decreased need to use the stack for parameter passing and temporary space are largely responsible for this increased difficulty.

1.7 Modification Tools

In addition to performing instrumentation, where the behavior of the original binary is not changed, Dyninst and its components allow modification of the binary. This modification can occur at the instruction level, at the CFG level, or even at the level of data layout on the stack. We present an example of each of these use cases.

CRAFT [18] is a tool that determines which double-precision values in a binary can best be replaced by single-precision, attempting to obtain the maximum performance benefit while ensuring that output accuracy remains within a user-specified tolerance. To do this, it replaces each double-precision instruction with a code sequence that performs the same operation in parallel in single and double precision, and then tracks the error introduced by conversion to single precision. Figure 1.5 illustrates this operation.

Bernat and Miller [6] demonstrated the use of Dyninst components to apply security patches at the binary level to a running process by matching a CFG fingerprint, constructing the code added by the patch in snippet form, and modifying the control flow of the binary appropriately. This application, unlike CRAFT, typically works by replacing blocks and edges as an entire subgraph of the CFG; Bernat and Miller's example patches the Apache HTTP server by wrapping a function call in an appropriate error checking and handling conditional. This CFG-based approach to binary modification does not rely on symbols or particular instruction patterns. This allows it to properly apply patches across binaries generated by a wide range of compilers, and to be robust against inlining of the location to be patched.

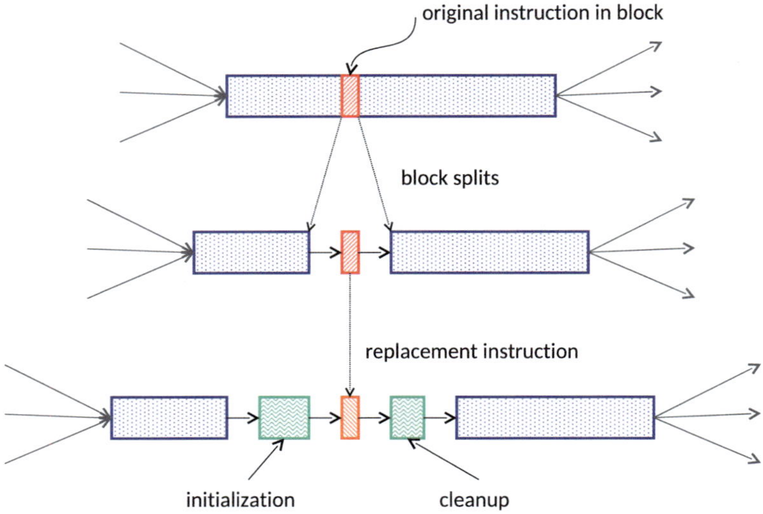

Fig. 1.5 Replacing instructions in basic blocks with CRAFT [18]

Gember-Jacobson and Miller [13] implemented primitives within Dyninst that allow the modification of functions' stack frames in well-specified manners: insertion and removal of space, and exchanging two local variables within the same contiguous stack region. This work does not alter the control flow of the binary at all; its purpose is solely to affect the data layout of the stack. In addition to the modifications that can be expressed purely in terms of insertion, removal, and exchange, they provide implementations for inserting stack canaries into functions and randomizing the order of local variables on the stack. Unlike the previous two examples, which altered the control flow graph of the program, this work modifies the data flow graph of the program while holding control flow constant.

1.8 Future Work

Dyninst and MRNet have become projects with a broad base of contributors and ongoing development. As we deconstructed Dyninst into smaller tool kits, we refined which complexities are actually necessary, and refined our abstractions to better match what users need. In particular, the deconstruction of Dyninst has shown us that Dyninst components may be used in a far broader set of applications than we initially expected.

In Dyninst, we plan to add full support for ARM64/Linux, add support for 64-bit Windows, and add support for Windows binary rewriting in the near term. We are also continually working to support new high-performance computing environments. In MRNet, we plan to implement a zero-copy interface that will improve performance.

Both Dyninst and MRNet are available via anonymous git checkout from http://git.dyninst.org. The Dyninst mailing list is dyninst-api@cs.wisc.edu. The MRNet mailing list is mrnet@cs.wisc.edu. Contributions, questions, and feature requests are always welcome.

Acknowledgments This work is supported in part by Department of Energy grant DE-SC0010474; National Science Foundation Cyber Infrastructure grants OCI-1234408 and OCI-1032341; and Department of Homeland Security under Air Force Research Lab contract FA8750-12-2-0289. The authors would also like to thank the many previous developers and users of Dyninst and MRNet.

References

1. Adhianto, L., Banerjee, S., Fagan, M., Krentel, M., Marin, G., Mellor-Crummey, J., Tallent, N.R.: HPCToolkit: tools for performance analysis of optimized parallel programs. Concurr. Comput.: Pract. Exp. **22**(6), 685–701 (2010)
2. Ahn, D.H., De Supinski, B.R., Laguna, I., Lee, G.L., Liblit, B., Miller, B.P., Schulz, M.: Scalable temporal order analysis for large scale debugging. In: Proceedings of the Conference on High Performance Computing Networking, Storage and Analysis (SC09). ACM, Portland, Oregon, November 2009
3. Arnold, D.C., Ahn, D.H., De Supinski, B.R., Lee, G.L., Miller, B.P., Schulz, M.: Stack trace analysis for large scale debugging. In: IEEE International Parallel and Distributed Processing Symposium, 2007 (IPDPS 2007). IEEE, Long Beach, California, March 2007
4. Bao, T., Burket, J., Woo, M., Turner, R., Brumley, D.: BYTEWEIGHT: Learning to recognize functions in binary code. In: 23rd USENIX Conference on Security Symposium (SEC). San Diego, California, August 2014
5. Bernat, A.R., Miller, B.P.: Anywhere, any-time binary instrumentation. In: Proceedings of the 10th ACM SIGPLAN-SIGSOFT workshop on Program analysis for software tools (PASTE). ACM, Szeged, Hungary, September 2011
6. Bernat, A.R., Miller, B.P.: Structured binary editing with a CFG transformation algebra. In: 2012 19th Working Conference on Reverse Engineering (WCRE). IEEE, Kingston, Ontario, October 2012
7. Buck, B., Hollingsworth, J.K.: An API for runtime code patching. Int. J. High Perform. Comput. Appl. **14**(4), 317–329 (2000)
8. Buntinas, D., Bosilca, G., Graham, R.L., Vallée, G., Watson, G.R.: A scalable tools communications infrastructure. In: 22nd International Symposium on High Performance Computing Systems and Applications, 2008 (HPCS 2008). IEEE, Ottawa, Ontario, April 2008
9. Cray, Inc.: Cray Programming Environment User's Guide. Cray, Inc. (2014)
10. Dinh, M.N., Abramson, D., Chao, J., DeRose, L., Moench, B., Gontarek, A.: Supporting relative debugging for large-scale UPC programs. Procedia Comput. Sci. **29**, 1491–1503 (2014)
11. DWARF Standards Committee: The DWARF Debugging Standard, version 4. http://dwarfstd.org (2013)
12. Eigler, F.C., Red Hat, Inc.: Problem solving with SystemTap. In: Proceedings of the Ottawa Linux Symposium. Citeseer, Ottawa, Ontario, July 2006
13. Gember-Jacobson, E.R., Miller, B.: Performing stack frame modifications on binary code. Technical report, Computer Sciences Department, University of Wisconsin, Madison (2015)
14. Harris, L., Miller, B.: Practical analysis of stripped binary code. ACM SIGARCH Comput. Archit. News **33**(5), 63–68 (2005)
15. Jensen, N.B., Karlsson, S., Quarfot Nielsen, N., Lee, G.L., Ahn, D.H., Legendre, M., Schulz, M.: Dysectapi: Scalable prescriptive debugging. In: International Conference for High Performance Computing, Networking, Storage and Analysis (SC14). New Orleans, Louisiana, November 2014

16. Jin, C., Abramson, D., Dinh, M.N., Gontarek, A., Moench, R., DeRose, L.: A scalable parallel debugging library with pluggable communication protocols. In: 2012 12th IEEE/ACM International Symposium on Cluster, Cloud and Grid Computing (CCGrid). IEEE, Ottawa, Ontario, May 2012
17. Laguna, I., Ahn, D.H., de Supinski, B.R., Gamblin, T., Lee, G.L., Schulz, M., Bagchi, S., Kulkarni, M., Zhou, B., Qin, F.: Debugging high-performance computing applications at massive scales. Commun. ACM **58**(9), 72–81 (2015)
18. Lam, M.O., Hollingsworth, J.K., de Supinski, B.R., LeGendre, M.P.: Automatically adapting programs for mixed-precision floating-point computation. In: Proceedings of the 27th International ACM Conference on International Conference on Supercomputing (SC13). ACM, Denver, Colorado, November 2013
19. Lee, G.L., Ahn, D.H., Arnold, D.C., De Supinski, B.R., Legendre, M., Miller, B.P., Schulz, M., Liblit, B.: Lessons learned at 208k: towards debugging millions of cores. In: International Conference for High Performance Computing, Networking, Storage and Analysis, 2008. (SC08). IEEE, Austin, Texas, November 2008
20. Llort, G., Servat, H.: Extrae. Barcelona Supercomputer Center. https://www.bsc.es/computer-sciences/extrae (2015)
21. Meng, X., Miller, B.: Binary code is not easy. Technical report, Computer Sciences Department, University of Wisconsin, Madison (2015)
22. Miller, B.P., Roth, P., DelSignore, J.: A path to operating system and runtime support for extreme scale tools. Technical report, TotalView Technologies LLC (2012)
23. Mußler, J., Lorenz, D., Wolf, F.: Reducing the overhead of direct application instrumentation using prior static analysis. In: Proceedings of the 17th International Conference on Parallel Processing-Volume Part I (Euro-Par 2011). Springer, Bordeaux, France, September 2011
24. Ravipati, G., Bernat, A.R., Rosenblum, N., Miller, B.P., Hollingsworth, J.K.: Toward the deconstruction of Dyninst. Technical report, Computer Sciences Department, University of Wisconsin, Madison. ftp://ftp.cs.wisc.edu/paradyn/papers/Ravipati07SymtabAPI.pdf (2007)
25. Rosenblum, N., Zhu, X., Miller, B.P., Hunt, K.: Learning to analyze binary computer code. In: 23rd National Conference on Artificial Intelligence (AAAI). AAAI Press, Chicago, Illinois, July 2008
26. Roth, P.C., Arnold, D.C., Miller, B.P.: MRNet: A software-based multicast/reduction network for scalable tools. In: Proceedings of the 2003 ACM/IEEE Conference on Supercomputing (SC03). ACM, Phoenix, Arizona, November 2003
27. Rutar, N., Hollingsworth, J.K.: Assigning blame: Mapping performance to high level parallel programming abstractions. In: Sips, H., Epema, D., Lin, H.X. (eds.) Euro-Par 2009 Parallel Processing. Lecture Notes in Computer Science, vol. 5704. Springer, Berlin, Heidelberg, Delft, The Netherlands, August 2009
28. Schulz, M., Ahn, D., Bernat, A., de Supinski, B.R., Ko, S.Y., Lee, G., Rountree, B.: Scalable dynamic binary instrumentation for Blue Gene/L. ACM SIGARCH Comput. Archit. News **33**(5), 9–14 (2005)
29. Schulz, M., Galarowicz, J., Maghrak, D., Hachfeld, W., Montoya, D., Cranford, S.: Open|SpeedShop: an open source infrastructure for parallel performance analysis. Sci. Program. **16**(2–3), 105–121 (2008)
30. Schwarz, B., Debray, S., Andrews, G.: Disassembly of executable code revisited. In: Proceedings of the Ninth Working Conference on Reverse Engineering (WCRE'02). IEEE Computer Society, Richmond, Virginia, October 2002
31. Shende, S.S., Malony, A.D., Morris, A.: Improving the scalability of performance evaluation tools. In: Proceedings of the 10th International Conference on Applied Parallel and Scientific Computing-Volume 2 (PARA 2010). Springer, Reykjavik, Iceland, June 2010
32. Shin, E.C.R., Song, D., Moazzezi, R.: Recognizing functions in binaries with neural networks. In: 24th USENIX Conference on Security Symposium (SEC). USENIX Association, Washington, D.C., August 2015
33. Weiser, M.: Program slicing. In: Proceedings of the 5th International Conference on Software Engineering (ICSE). IEEE Press, San Diego, California, March 1981

Chapter 2
Validation of Hardware Events for Successful Performance Pattern Identification in High Performance Computing

Thomas Röhl, Jan Eitzinger, Georg Hager and Gerhard Wellein

Abstract Hardware performance monitoring (HPM) is a crucial ingredient of performance analysis tools. While there are interfaces like LIKWID, PAPI or the kernel interface perf_event which provide HPM access with some additional features, many higher level tools combine event counts with results retrieved from other sources like function call traces to derive (semi-)automatic performance advice. However, although HPM is available for x86 systems since the early 90s, only a small subset of the HPM features is used in practice. Performance patterns provide a more comprehensive approach, enabling the identification of various performance-limiting effects. Patterns address issues like bandwidth saturation, load imbalance, non-local data access in ccNUMA systems, or false sharing of cache lines. This work defines HPM event sets that are best suited to identify a selection of performance patterns on the Intel Haswell processor. We validate the chosen event sets for accuracy in order to arrive at a reliable pattern detection mechanism and point out shortcomings that cannot be easily circumvented due to bugs or limitations in the hardware.

2.1 Introduction and Related Work

Hardware performance monitoring (HPM) was introduced for the x86 architecture with the Intel Pentium in 1993 [15]. Since that time, HPM gained more and more attention in the computer science community and consequently a lot of HPM related tools were developed. Some provide basic access to the HPM registers with some additional features like LIKWID [17], PAPI [12] or the kernel interface perf_event [4]. Furthermore, some higher level analysis tools gather additional information by combining the HPM counts with application level traces. Popular representatives of that analysis method are HPCToolkit [1], PerfSuite [10], Open|Speedshop [16] or Scalasca [3]. The intention of these tools is to advise the application developer with educated optimization hints. To this end, the tool

T. Röhl (✉) · J. Eitzinger · G. Hager · G. Wellein
Erlangen Regional Computing Center (RRZE), University of Erlangen-Nuremberg, Erlangen, Germany
e-mail: thomas.roehl@fau.de

A. Knüpfer et al. (eds.), *Tools for High Performance Computing 2015*,
DOI 10.1007/978-3-319-39589-0_2

developers use performance metrics that represent a possible performance limitation, such as saturated memory bandwidth or instructions paths. The hardware metrics may be combined with information on the application level, e.g. scaling characteristics, dependence of performance on the problem size, or static code analysis, to arrive at a *signature*. A performance signature then points towards one or more *performance patterns*, as described in [19] and refined in [18]. The purpose of the patterns concept is to facilitate the identification of performance-limiting bottlenecks.

C. Guillen uses in [5] the term *execution properties* instead of performance pattern. She defines execution properties as a set of values gathered by monitoring and related thresholds. The properties are arranged in decision trees for compute- and memory-bound applications as well as trees related to I/O and other resources. This enables either a guided selection of the analysis steps to further identify performance limitation or automatic tool-based analysis. Based on the path in the decision tree, suggestions are given for what to look for in the application code. A combination of the structured performance engineering process in [18] with the decision trees in [5] defines a good basis for (partially) automated performance analysis tools.

One main problem with HPM is that none of the main vendors for x86 processors guarantees event counts to be accurate or deterministic. Although many HPM interfaces exist, only little research has been done on validating the hardware performance events. However, users tend to trust the returned HPM counts and use them for decisions about code optimization. One should be aware that HPM measurements are only guideposts until the HPM events are known to have guaranteed behavior. Moreover, analytic performance models can only be validated if this is the case.

The most extensive event validation analysis was done by Weaver et al. [20] using a self-written assembly validation code. They test determinism and overcounting for the following events: retired instructions, retired branches, retired loads and stores as well as retired floating-point operations including scalar, packed, and vectorized instructions. For validating the measurements the dynamic binary instrumentation tool *Pin* [11] was used. The main target of that work was not to identify the right events needed to construct accurate performance metrics but to find the sources of non-determinism and over/undercounting. It gives hints on how to reduce over- or undercounting and identify deterministic events for a set of architectures.

D. Zaparanuks et al. [21] determined the error of retired instructions and CPU cycle counts with two microbenchmarks. Since the work was released before the perf_event interface [4] was available for PAPI, they tested the deprecated interfaces perfmon2 [2] and perfctr [13] as the basis for PAPI. They use an "empty" microbenchmark to define a default error using different counter access methods. For subsequent measurements they use a simple loop kernel with configurable iterations, define a model for the code and compare the measurement results to the model. Moreover, they test whether the errors change for increasing measurement duration and for a varying number of programmed counter registers. Finally, they give suggestions which back-end should be used with which counter access pattern to get the most accurate results.

In the remainder of this section we recommend HPM event sets and related derived metrics that represent the signature of prototypical examples picked out of the performance patterns defined in [18]. In the following sections the accuracy of the chosen HPM events and their derived metrics is validated. Our work can be seen as a recommendation for tool developers which event sets match the selected performance patterns in the best way and how reliable they are.

2.2 Identification of Signatures for Performance Patterns

Performance patterns help to identify possible performance problems in an application. The measurement of HPM events is one part of the pattern's signature. There are patterns that can be identified by HPM measurements alone, but commonly more information is required, e.g., scaling behavior or behavior with different data set sizes. Of course, some knowledge about the micro-architecture is also required to select the proper event sets for HPM as well as to determine the capabilities of the system. For x86 systems, HPM is not part of the instruction set architecture (ISA), thus besides a few events spanning multiple micro-architectures, each processor generation defines its own list of HPM events. Here we choose the Intel Haswell EP platform (E5-2695 v3) for HPM event selection and verification. The general approach can certainly be applied to other architectures.

In order to decide which measurement results are good or bad, the characteristics of the system must be known. C. Guillen established thresholds in [5] with four different approaches: hardware characteristics, expert knowledge about hardware behavior and performance optimization, benchmarks and statistics. With decision trees but without source code knowledge it is possible to give some loose hints how to further tune the code. With additional information about the software code and run time behavior, the list of hints could be further reduced.

The present work is intended to be a referral for which HPM events provide the best information to specify the signatures of the selected performance patterns. The patterns target different behaviors of an application and/or the hardware and therefore are classified in three groups: *bottlenecks*, *hazards* and *work-related* patterns. The whole list of performance patterns with corresponding event sets for the Intel Haswell EP micro-architecture can be found at [14]. For brevity we restrict ourselves to three patterns: *bandwidth saturation*, *load imbalance* and *false sharing of cache lines*. For each pattern, we list possible signatures and shortcomings concerning the coverage of a pattern by the event set. The analysis method is comparable to the one of D. Zaparanuks et al. [21] but uses a set of seven assembly benchmarks and synthetic higher level benchmark codes that represent often used algorithms in scientific applications. But instead of comparing the raw results, we use derived metrics, combining multiple counter values, for comparison as these metric results are commonly more interesting for tool users.

2.2.1 Bandwidth Saturation

A very common bottleneck is bandwidth saturation in the memory hierarchy, notably at the memory interface but also in the L3 cache on earlier Intel designs. Proper identification of this pattern requires an accurate measurement of the data volume, i.e., the number of transferred cache lines between memory hierarchy levels. From data volume and run time one can compute transfer bandwidths, which can then be compared with measured or theoretical upper limits.

Starting with the Intel Nehalem architecture, Intel separates a CPU socket in two components, the core and the uncore. The core embodies the CPU cores and the L1 and L2 caches. The uncore covers the L3 cache as well as all attached components like memory controllers or the Intel QPI socket interconnect. The transferred data volume to/from memory can be monitored at two distinct uncore components. A CPU socket in an Intel Haswell EP machine has at most two memory controllers (iMC) in the uncore, each providing up to four memory channels. The other component is the Home Agent (HA) which is responsible for the protocol side of memory interactions.

Starting with the Intel Sandy Bridge micro-architecture, the L3 cache is segmented, with one segment per core. Still one core can make use of all segments. The data transfer volume between the L2 and L3 caches can be monitored in two different ways: One may either count the cache lines that are requested and written back by the L2 cache, or the lookups for data reads and victimized cache lines that enter the L3 cache segments. It is recommended to use the L2-related HPM events because the L3 cache is triggered by many components besides the L2 caches. Moreover, the Intel Haswell EP architecture has up to 18 L3 cache segments which all need to be configured separately. Bandwidth bottlenecks between L1 and L2 cache or L1 and registers are seldom and thus ignored in this pattern.

2.2.2 Load Imbalance

The main characterization of this pattern is that different threads have to process different working sets between synchronization points. For data-centric workloads the data volume transferred between the L1 and L2 caches for each thread may be an indicator: since the working sets have different sizes, it is likely that smaller working sets also require less data. However, the assumption that working set size is related to transferred cache lines is not expressive enough to fully identify the pattern, since the amount of required data could be the same for each thread while the amount of in-core instructions differs. Retired instructions, on the other hand, are just as unreliable as data transfers because parallelization overhead often comprises spin-waiting loops that cause abundant instructions without doing "work." Therefore, for better classification, it is desirable to count "useful" instructions that perform the actual work the application has to do. None of the two x86 vendors provides features to filter the instruction stream and count only specific instructions in a sufficiently flexible way. Moreover, the offered hardware events are not sufficient to overcome this shortcoming by covering most "useful" instructions like scalar/packed floating-point operations, SSE driven calculations or string related operations. Nevertheless, filtering on some instruction groups works for Intel Haswell systems, such as long-latency instructions (div, sqrt, …) or AVX instructions. Consequently, it is recommended to measure the work instructions if possible but also the data transfers can give a first insight.

2.2.3 False Cache Line Sharing

False cache line sharing occurs when multiple cores access the same cache line while at least one is writing to it. The performance pattern thus has to identify bouncing cache lines between multiple caches. There are codes that require true cache line sharing, like producer/consumer codes, but we are referring to common HPC codes where cache line sharing should be as minimal as possible. In general, the detection of false cache line sharing is very hard when restricting the analysis space only to hardware performance measurements. The Intel Haswell micro-architecture offers two options for counting cache line transfers between private caches: There are L3 cache related μOPs events for intra- and inter-socket transfers, but the HPM event for intra-socket movement may undercount with SMT enabled by as much as 40 % according to erratum HSW150 in [8]. The alternative is the *offcore response unit*. By setting the corresponding filter bits, the L3 hits with hitm snoops (hit a modified cache line) to other caches on the socket and the L3 misses with hitm snoops to remote sockets can be counted. The specification update [8] also lists an erratum for the offcore response unit (HSW149) but the required filter options for shared cache lines are not mentioned in it. There are no HPM events to count the transfers of shared cache lines at the L2 cache. In order to clearly identify whether a code triggers true or false cache line sharing, further information like source code analysis is required.

2.3 Useful Event Sets

Table 2.1 defines a range of HPM event sets that are best suitable for the described performance patterns regarding the HPM capabilities of the Intel Haswell EP platform. The assignment of HPM events for the pattern signatures is based on the Intel documentation [6, 9]. Some events are not mentioned in the default documentation; they are taken from Intel's performance monitoring database [7]. Although the events were selected with due care, there is no official guarantee for the accuracy of the counts by the manufacturer. The sheer amount of performance monitoring related errata for the Intel Haswell EP architecture [8] reduces the confidence even further. But this encourages us even more to validate the chosen event sets in order to provide tool developers and users a reliable basis for their performance analysis.

Table 2.1 Desired events and available events for three performance patterns on the Intel Haswell EP micro-architecture

Pattern	Desired events	Available events
Bandwidth saturation	Data volume transferred to/from memory from/to the last level cache; data volume transferred between L2 and L3 cache	iMC:UNC_M_CAS_COUNT.RD, iMC:UNC_M_CAS_COUNT.WR, HA:UNC_H_IMC_READS.NORMAL, HA:UNC_H_BYPASS_IMC.TAKEN, HA:UNC_H_IMC_WRITES.ALL, L2_LINES_IN.ALL, L2_TRANS.L2_WB, CBOX:LLC_LOOKUP.DATA_READ, CBOX:LLC_VICTIMS.M_STATE
Load imbalance	Data volume transferred at all cache levels; number of "useful" instructions	L1D.REPLACEMENT, L2_TRANS.L1D_WB, L2_LINES_IN.ALL, L2_TRANS.L2_WB, AVX_INSTS.CALC, ARITH.DIVIDER_UOPS
False sharing of cache lines	All transfers of shared cache lines for the L2 and L3 cache; all transfers of shared cache lines between the last level caches of different CPU sockets	MEM_LOAD_UOPS_L3_HIT_RETIRED.XSNP_HITM, MEM_LOAD_UOPS_L3_MISS_RETIRED.REMOTE_HITM, OFFCORE_RESPONSE:LLC_HIT:HITM_OTHER_CORE, OFFCORE_RESPONSE:LLC_MISS:REMOTE_HITM

A complete list can be found at [14]

2.4 Validation of Performance Patterns

Many performance analysis tools use the HPM features of the system as their main source of information about a running program. They assume event counts to be correct, and some even generate automated advice for the developer. Previous research in the field of HPM validation focuses on singular events like retired instructions but does not verify the results for other metrics that are essential for identifying performance patterns. Proper verification requires the creation of benchmark code that has well-defined and thoroughly understood performance features and, thus, predictable event counts. Since optimizing compilers can mutilate the high level code, the feasible solutions are either to write assembly benchmarks or to perform code analysis of the assembly code created by the compiler.

The LIKWID tool suite [17] includes the `likwid-bench` microbenchmarking framework, which provides a set of assembly language kernels. They cover a variety of streaming access schemes. In addition the user can extend the framework by writing new assembly code loop bodies. `likwid-bench` takes care of loop counting, thread parallelism, thread placement, ccNUMA page placement and performance (and bandwidth) measurement. It does not, however, perform hardware event counting. For the HPM measurements we thus use `likwid-perfctr`, which is also a part of the LIKWID suite. It uses a simple command line interface but provides a comprehensive set of features for the users. `Likwid-perfctr` supports almost all interesting core and uncore events for the supported CPU types. In order to relieve the user from having to deal with raw event counts, it supports *performance groups*, which combine often used event sets and corresponding formulas for computing derived metrics (e.g., bandwidths or FLOP rates). Moreover, `likwid-perfctr` provides a *Marker API* to instrument the source code and restrict measurements to certain code regions. `Likwid-bench` already includes the calls to the Marker API in order to measure only the compute kernel. We have to manually correct some of the results of `likwid-bench` to represent the obvious and hidden data traffic (mostly write-allocate transfers) that may be measured with `likwid-perfctr`.

The first performance pattern for the analysis is the bandwidth saturation pattern. For this purpose, `likwid-perfctr` already provides three performance groups called L2, L3 and MEM [17]. A separate performance group was created to measure the traffic traversing the HA. Based on the raw counts, the groups define derived metrics for data volume and bandwidth. For simplicity we use the derived metric of total bandwidth for comparison as it both includes the data volume in both directions and the run time. In Fig. 2.1 the average, minimal and maximal errors of 100 runs

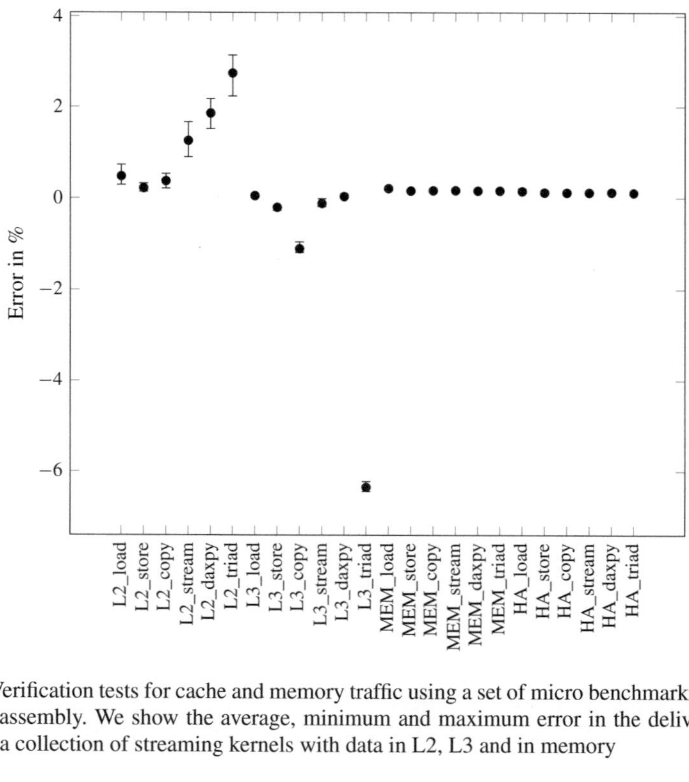

Fig. 2.1 Verification tests for cache and memory traffic using a set of micro benchmarking kernels written in assembly. We show the average, minimum and maximum error in the delivered HPM counts for a collection of streaming kernels with data in L2, L3 and in memory

with respect to the exact bandwidth results are presented for seven streaming kernels and data in L2 cache, L3 cache and memory. The locality of the data in the caching hierarchy is ensured by streaming accesses to the vectors fitting only in the relevant hierarchy level. The first two kernels (*load* and *store*) perform pure loading and storing of data to/from the CPU core to the selected cache level or the memory. A combination of both is applied in the *copy* test. The last three tests are related to scientific computing and well understood. They range from the linear combination of two vectors called *daxpy* calculating $A[i] = B[i] \cdot c + A[i]$, a *stream* triad with formula $A[i] = B[i] \cdot c + C[i]$ to a vector *triad* computing $A[i] = B[i] \cdot C[i] + D[i]$.

The next pattern we look at is load imbalance. Since load imbalance requires a notion of "useful work" we have to find a way to measure floating-point operations. Unfortunately, the Intel Haswell architecture lacks HPM events to fully represent FLOP/s. For the Intel Haswell architecture, Intel has documented a HPM event `AVX_INSTS.ALL` (Event 0xC6, Umask 0x07) which captures all AVX instructions including data movement and calculations [7]. With the help of `likwid-bench` we could further refine the event to count loads (Umask 0x01), stores (Umask 0x02) and calculations (Umask 0x04) separately. Consequently, the FLOP/s performed with AVX operations can be counted. All performance patterns that require the filtering of the instruction stream for specific instructions can use the event `AVX_INSTS.CALC` for floating-point operations using the AVX vectorization extension. Due to its importance, the event is verified using the `likwid-bench` utility with assembly benchmarks that are based on AVX instructions only. Note that the use of these specific Umasks is an undocumented feature and may change with processor generations or even mask revisions. Moreover, we have found no way to count SSE or scalar floating-point instructions.

Figure 2.2 shows the minimum, maximum and average error for measuring AVX FLOP/s. The average error for all tests is below 0.07 %. As the maximal error is 0.16 % the event can be seen as sufficiently accurate for pure AVX code. Using the counter with non-AVX codes always returns 0.

Coming back to performance patterns, we now verify the load imbalance pattern using an upper triangular matrix vector multiplication code running with two threads. Since the accuracy of the cache and memory traffic related HPM events have been verified already, we use the only available floating-point operation related event `AVX_INSTS.CALC`. There is one shortcoming worth noting: If the code contains half-wide loads, the HPM event shows overcounting. The compiler frequently uses half-wide loads to reduce the probability of "split loads," i.e., AVX loads that cross a cache line boundary if 32-byte alignment cannot be guaranteed. Experiments have shown that the event `AVX_INSTS.CALC` includes the `vinsertf128` instruction as a calculation operation. In order to get reliable results, split AVX loads should be avoided. This is not a problem with `likwid-bench` as no compiler is involved and the generated assembly code is under full control. The upper triangular matrix is split so that each of the two threads operates on half of the matrix. The matrix has a size of 8192×8192 and the multiplication is performed 1000 times. The first thread processes the top rows with totally 25,167,872 elements, while the second

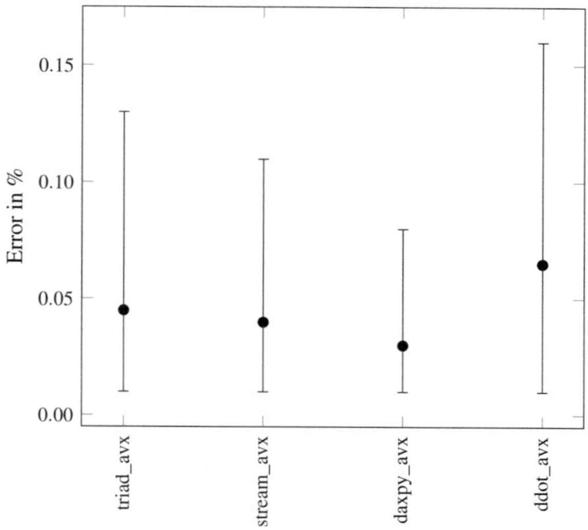

Fig. 2.2 Verification tests for the AVX floating point event using a set of microbenchmarking kernels with pure AVX code

Table 2.2 Verification of the load imbalance pattern using an upper triangular matrix vector multiplication code

Event/Metric	Thread 0	Thread 1	Ratio	Error
Process elements	25167872	8390656	3 : 1	
AVX floating point ops	1.26E10	4.21E09	2.991 : 1	0.29 %
L2 data volume [GByte]	406.28	115.34	3.52 : 1	17.42 %
L3 data volume [GByte]	203.06	69.74	2.912 : 1	2.94 %
Memory data volume [GByte]	112.97	37.33	3.026 : 1	0.88 %

one works on the remaining 8,390,656 elements. This distribution results in a work load imbalance for the threads of 3 : 1.

Table 2.2 lists the verification data for the code. The AVX calculation instruction count fits to a high degree the work load ratio of 3 : 1. The L2 data volume has the highest error, mainly caused by repeatedly fetching the input and output vector not included in the work load balance model. This behavior also occurs for the L3 and memory data volume but to a lesser extent as the cache lines of the input vector commonly stay in the caches. In order to get the memory data volume per core, the offcore response unit was used.

The false sharing of cache lines pattern is difficult to verify as it is not easy to write code that shows a predictable number of inter-core cache line transfers. A minimal amount of shared cache lines exist in almost every code thus HPM results unequal zero cannot be accepted as clear signature. To measure the behavior, a producer and

Table 2.3 Verification tests for false sharing of cache lines using a producer/consumer code

Amount of shared cache lines per step	Transferred cache lines according to model	Avg. amount of intra-socket transferred shared cache lines	Error [%]	Avg. amount of inter-socket transferred shared cache lines	Error [%]
2	200	328.4	64.2	404.2	102.1
4	400	568.3	42.1	607.7	51.9
8	800	897.0	12.1	985.86	23.2
16	1600	1714.2	7.1	1777.1	11.1
32	3200	2893.4	−9.6	2595.6	−18.9
64	6400	5570.8	−13.0	3512.3	−45.1
128	12,800	7350.7	−42.6	6124.9	−52.2
256	25,600	8995.3	−64.9	11471.1	−55.2
512	51,200	18224.3	−64.4	22608.3	−55.8
1024	102,400	55124.5	−46.2	45814.3	−55.3

The producer and consumer thread are located on the same CPU socket

consumer code was written, thus we verify the amount of falsely shared cache lines by using a true sharing cache line code. The producer writes to a consecutive range of memory that is read afterwards by the consumer. In the next iteration the producer uses the subsequent range of memory to avoid invalidation traffic. The memory range is aligned so that a fixed amount of cache lines is used in every step. The producer and consumer perform 100 iterations in each of the 100 runs. For synchronizing the two threads, a simple busy-waiting loop spins on a shared variable with long enough sleep times to avoid high access traffic for the synchronization variable. When using pthread conditions and a mutex lock instead, the measured values are completely unstable.

Table 2.3 shows the measurements for HPM events fitting best to the traffic caused by false sharing of cache lines. The table lists the amount of cache lines that are written by the producer thread. Since the consumer reads all these lines, the amount of transferred cache lines should be in the same range. The measurements using the events in Table 2.1 show a big discrepancy between the counts in the model and the measured transfers. For small counts of transferred cache lines, the results are likely to be distorted by the shared synchronization variable, but the accuracy should improve with increasing transfer sizes. Since the erratum HSW150 in [8] states an undercounting by as much as 40 %, the intra-socket measurements could be too low. But even when scaling up the measurements the HPM event for intra-socket cache line sharing is not accurate.

For the inter-socket false sharing, the threads are distributed over the two CPU sockets in the system. The results in Table 2.3 show similar behavior as in the intra-socket case. The HPM events for cache line sharing provide a qualitative classification for the performance pattern's signature but no quantitative one. The problem is mainly to define a threshold for the false-sharing rate of the system and application. Further research is required to create suitable signature for this performance pattern.

2.5 Conclusion

The performance patterns defined in [18] provide a comprehensive collection for analyzing possible performance degradation on the node level. They address possible hardware bottlenecks as well as typical inefficiencies in parallel programming. We have listed suitable event sets to identify the bandwidth saturation, load imbalance, and false sharing patterns with HPM on the Intel Haswell architecture. Unfortunately the hardware does not provide all required events, such as, e.g., scalar/packed floating-point operations, or they are not accurate enough like, e.g., the sharing of cache lines at the L3 level. Moreover, a more fine-grained and correct filtering of instructions would be helpful for pattern-based performance analysis.

Using a selection of streaming loop kernels we found the error for the bandwidth-related events to be small on average ($-1\,\% \ldots +2\,\%$), with a maximum under-counting of about $-6\,\%$ for the L3 traffic. The load imbalance pattern was verified using an upper triangular matrix vector multiplication. Although the error for the L1 to L2 cache traffic is above 15 %, the results reflect the correct load imbalance of roughly $3:1$, indicating the usefulness of the metrics. Moreover, we have managed to identify filtered events that can accurately count AVX floating-point operations under some conditions. FLOP/s and traffic data are complementary information for identifying load imbalance. The verification of the HPM signature for the false sharing pattern failed due to large deviations from the expected event counts for the two events used. More research is needed here to arrive at a useful procedure, especially for distinguishing unwanted false cache line sharing from traffic caused by intended updates.

The remaining patterns defined in [18] need to be verified as well to provide a well-defined HPM analysis method for performance patterns ready to be included in performance analysis tools. We provide continuously updated information about suitable events for pattern identification in the Wiki on the LIKWID website.[1]

Acknowledgments Parts of this work were funded by the German Federal Ministry of Research and Education (BMBF) under Grant Number 01IH13009.

References

1. Adhianto, L., Banerjee, S., Fagan, M., Krentel, M., Marin, G., Mellor-Crummey, J., Tallent, N.R.: HPCToolkit: Tools for performance analysis of optimized parallel programs. Concurr. Comput.: Pract. Exp.D **22**(6), 685–701 (2010)

[1] https://github.com/RRZE-HPC/likwid.

2. Eranian, S.: Perfmon2: a flexible performance monitoring interface for Linux. In: Ottawa Linux Symposium, pp. 269–288, Citeseer (2006)
3. Geimer, M., Wolf, F., Wylie, B.J., Ábrahám, E., Becker, D., Mohr, B.: The Scalasca performance toolset architecture. Concurr. Comput.: Pract. Exp. **22**(6), 702–719 (2010)
4. Gleixner, T., Molnar, I.: Linux 2.6.32: perf_event.h. http://lwn.net/Articles/310260/ (2008)
5. Guillen, C.: Knowledge-based performance monitoring for large scale HPC architectures. Dissertation p. http://mediatum.ub.tum.de/?id=1237547 (2015)
6. Intel: Intel 64 and IA-32 Architectures Software Developer Manuals. http://www.intel.com/content/www/us/en/processors/architectures-software-developer-manuals.html (2015)
7. Intel: Intel Open Source Technology Center for PerfMon. https://download.01.org/perfmon/ (2015)
8. Intel: Intel Xeon Processor E3-1200 v3 Product Family Specification Update. http://www.intel.com/content/dam/www/public/us/en/documents/specification-updates/xeon-e3-1200v3-spec-update.pdf (2015)
9. Intel: Intel Xeon Processor E5 v3 Family Uncore Performance Monitoring. https://www-ssl.intel.com/content/dam/www/public/us/en/zip/xeon-e5-v3-uncore-performance-monitoring.zip (2015)
10. Kufrin, R.: Perfsuite: An accessible, open source performance analysis environment for linux. In: 6th International Conference on Linux Clusters: The HPC Revolution, vol. 151, p. 05. Citeseer (2005)
11. Luk, C.K., Cohn, R., Muth, R., Patil, H., Klauser, A., Lowney, G., Wallace, S., Reddi, V.J., Hazelwood, K.: Pin: Building customized program analysis tools with dynamic instrumentation. SIGPLAN Not. **40**(6), 190–200 (2005). http://doi.acm.org/10.1145/1064978.1065034
12. Mucci, P.J., Browne, S., Deane, C., Ho, G.: PAPI: A portable interface to hardware performance counters. In: Proceedings of the Department of Defense HPCMP Users Group Conference. pp. 7–10 (1999)
13. Pettersson, M.: Linux x86 performance-monitoring counters driver (2003)
14. Roehl, T.: Performance patterns for the Intel Haswell EP/EN/EX architecture. https://github.com/RRZE-HPC/likwid/wiki/PatternsHaswellEP (2015)
15. Ryan, B.: Inside the Pentium. BYTE Mag. **18**(6), 102–104 (1993)
16. Schulz, M., Galarowicz, J., Maghrak, D., Hachfeld, W., Montoya, D., Cranford, S.: Open|SpeedShop: An open source infrastructure for parallel performance analysis. Sci. Prog. **16**(2–3), 105–121 (2008)
17. Treibig, J., Hager, G., Wellein, G.: LIKWID: A lightweight performance-oriented tool suite for x86 multicore environments. In: Proceedings of PSTI2010, the First International Workshop on Parallel Software Tools and Tool Infrastructures. San Diego, CA (2010)
18. Treibig, J., Hager, G., Wellein, G.: Pattern driven node level performance engineering. http://sc13.supercomputing.org/sites/default/files/PostersArchive/tech_posters/post254s2-file2.pdf (2013), sC13 poster
19. Treibig, J., Hager, G., Wellein, G.: Performance patterns and hardware metrics on modern multicore processors: Best practices for performance engineering. Euro-Par 2012: Parallel Processing Workshops. Lecture Notes in Computer Science, vol. 7640, pp. 451–460. Springer, Berlin (2013)
20. Weaver, V., Terpstra, D., Moore, S.: Non-determinism and overcount on modern hardware performance counter implementations. In: 2013 IEEE International Symposium on Performance Analysis of Systems and Software (ISPASS), pp. 215–224 (2013)
21. Zaparanuks, D., Jovic, M., Hauswirth, M.: Accuracy of performance counter measurements. In: IEEE International Symposium on Performance Analysis of Systems and Software, 2009. ISPASS 2009. pp. 23–32 (2009)

Chapter 3
Performance Optimization for the Trinity RNA-Seq Assembler

Michael Wagner, Ben Fulton and Robert Henschel

Abstract Utilizing the enormous computing resources of high performance computing systems is anything but a trivial task. Performance analysis tools are designed to assist developers in this challenging task by helping to understand the application behavior and identify critical performance issues. In this paper we share our efforts and experiences in analyzing and optimizing Trinity, a well-established framework for the de novo reconstruction of transcriptomes from RNA-seq reads. Thereby, we try to reflect all aspects of the ongoing performance engineering: the identification of optimization targets, the code improvements resulting in 22 % overall runtime reduction, as well as the challenges we encountered getting there.

3.1 Introduction

High performance computing (HPC) systems promise to provide enormous computational resources. But effectively utilizing the computational power of these systems requires increasing knowledge and effort. Along with efficient single thread performance and resource usage, developers must consider various parallel programming models such as message passing, threading and tasking, and architecture specific models like interfaces to incorporate GPUs. Appropriate development devices such as performance analysis tools are becoming increasingly important in utilizing the computational resources of today's HPC systems. They assist developers in two key aspects of program development: first, they help to analyze and understand the

M. Wagner (✉)
Barcelona Supercomputing Center, Barcelona, Spain
e-mail: michael.wagner@bsc.es

M. Wagner
Center for Information Services and High Performance Computing,
Technische Universität Dresden, Dresden, Germany

B. Fulton · R. Henschel
Scientific Applications and Performance Tuning Indiana University,
E. Tenth Street, Bloomington, IN 2709, USA

A. Knüpfer et al. (eds.), *Tools for High Performance Computing 2015*,
DOI 10.1007/978-3-319-39589-0_3

behavior of the applications on the HPC system and, second, they help in identifying critical performance issues.

In this paper we present our efforts to analyze and optimize the RNA-Seq assembler Trinity [4]. Trinity is a software tool that was developed for accurate de novo reconstruction of transcriptomes from RNA-Seq data. Early versions of the tool required a great deal of memory and performant hardware. As part of an ongoing process of performance improvement, we used Collectl [2], Score-P [8], and Vampir [7] to identify bottlenecks in the Trinity pipeline, with diverse causes including memory contention, suboptimal I/O, and streaming inefficiencies. Armed with this knowledge, we were able to introduce modifications resulting in a 22 % improvement in overall wall time.

In the following section we present the tool infrastructure that we used to gain insight into the application behavior and performance characteristics. In Sect. 3.3 we focus on the methods we used to understand Trinity's overall behavior and the behavior of the individual components. Furthermore, we will demonstrate the resulting optimizations in the Trinity pipeline. In Sect. 3.4 we discuss certain challenges and restrictions we encountered while using various tools. Finally, we summarize the presented work and draw conclusions.

3.2 Tool Infrastructure

To better understand the runtime behavior of Trinity, to identify targets for performance optimization, and also to analyze the performance we used state-of-the-art performance tools: the system performance monitor Collectl, the event-based trace collecter Score-P and the visual performance analyzer Vampir.

3.2.1 Collectl

Collectl is a popular performance monitoring tool that is able to track a wide variety of subsystems, including CPU, disk accesses, inodes, memory usage, network bandwidth, nfs, processes, quadrics, slabs, sockets, and TCP [2]. It is additionally popular with HPC administrators for its ability to monitor clusters and to track network and file systems such as InfiniBand and Lustre. Collectl works at a high level by sampling the system at intervals to determine the usage of each resource and logs the information to a file.

Collectl has long been incorporated into the Trinity pipeline to monitor various statistics at a coarse-grained level of detail. To minimize the effect on performance, Trinity runs Collectl at a sampling rate of five seconds, rather than the default one second rate, and only monitors applications launched by the current user.

We extracted statistics from the Collectl log generated by Trinity on RAM usage, CPU utilization, and I/O throughput, and created charts summarizing the use of each individual application in the Trinity pipeline (see Fig. 3.1). In order to determine the performance of each pipeline component, the totals were summed up regardless of whether the component consisted of an single, multi-threaded application, or multiple copies of an application running simultaneously. From these charts, we were able to coarsely assess the relative amount of time each component used, as well as how effectively it made use of available resources.

3.2.2 Score-P and Vampir

For a more detailed analysis we chose the state-of-the-art event trace monitor Score-P and the visual trace analyzer Vampir. Score-P is a joint measurement infrastructure for the analysis tools Vampir, Scalasca, Periscope, and TAU [8]. It incorporates the measurement functionality of these tools into a single infrastructure, which provides a maximum of convenience for users. The Score-P measurement infrastructure allows event tracing as well as profiling. It contains the code instrumentation functionality and performs the runtime data collection. For event tracing, Score-P uses the Open Trace Format 2 (OTF2) to store the event tracing data for a successive analysis [3]. The Open Trace Format 2 is a highly scalable, memory efficient event trace data format plus support library.

Vampir is a well-proven and widely used tool for event-based performance analysis in the high performance computing community [7]. The Vampir trace visualizer includes a scalable, distributed analysis architecture called VampirServer, which enables the scalable processing of both large amounts of trace data and large numbers of processing elements. It presents the tracing data in the form of timelines, displaying the active code region over time for each process along with summarized profile information, such as the amount of time spent in individual functions.

3.3 Analysis and Optimization

The starting point for the optimization was Trinity 2.0.6 [5] which already contains a number of previous optimization cycles [6]. Trinity 2.0.6 is a pipeline of up to 27 individual components in different programming and script languages, including C++, Java, Perl, and system binaries, which are invoked by the main Trinity perl script. The pipeline consists of three stages: first, *Inchworm* assembles RNA-seq data into sequence contigs, second, *Chrysalis* bundles the Inchworm contigs and constructs complete de Bruijn graphs for each cluster, and, third, *Butterfly* processes the individual graphs in parallel and computes the final assembly.

3.3.1 *Identification of Optimization Targets*

Due to the multicomponent structure of Trinity, many performance analysis tools which focus on a single binary were unsuitable to gain a general overview on the Trinity runtime behavior. To better understand the runtime behavior and to identify targets for optimization, we conducted a series of reference runs using Collectl to measure timings and resource utilization. Figure 3.1 depicts the initial performance of nine main components in Trinity 2.0.6, processing the 16.4 GiB reference data set of *Schizosaccharomyces Pombe*, a yeast, with 50 million base pairs on a 16-core node on the Karst cluster at Indiana University.

Based on the CPU utilization of the individual components we identified *Inchworm*, *Scaffold_iworm_contigs*, *Sort*, and *Butterfly* to run in serial or with insufficient parallel efficiency. *Inchworm* has already been targeted for a complete reimplementation using MPI in a different group and, therefore, was not selected as optimization target again [1]. The optimization of *Scaffold_iworm_contigs* is discussed in Sect. 3.3.2 and the optimization of *Sort* is highlighted in Sect. 3.3.3. The third stage of Trinity processing primarily involves *Butterfly*. An optimization of *Butterfly* would have implied a complete restructuring of the Trinity code, which was infeasible due to Trinity's modular and constantly evolving pipeline. Nevertheless, the second stage recursively calls the main Trinity script, and therefore this stage benefits from our other optimization efforts as each individual de Bruijn graph is processed.

In addition to the obvious optimization targets, we discovered an overhead of frequent forking and joining of parallel regions in *ReadsToTranscipts* marked by the sharp drops of parallel CPU utilization in the Collectl chart (Fig. 3.1). The resulting optimizations are discussed in Sect. 3.3.4.

While Collectl's CPU utilization displays insufficient multi-core usage it does not expose unbalanced parallel behavior, for instance, busy-waiting cores. Therefore, we analyzed the parallel scaling of the individual components to detect poor scaling components. Table 3.1 lists the parallel speedup of each component together

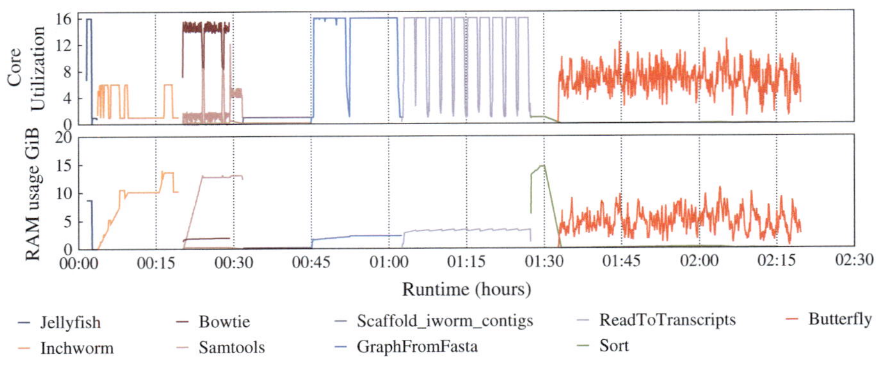

Fig. 3.1 Resource utilization of the original Trinity 2.0.6. version

Table 3.1 Parallel speedup of the main Trinity components together with their runtime share

Number of cores	Parallel speedup					Component runtime share (in %)				
	1	2	4	8	16	1	2	4	8	16
Trinity (total)	1.00	2.07	3.80	6.22	8.78	100.0	100.0	100.0	100.0	100.0
Jellyfish	1.00	1.84	3.23	5.20	7.34	1.4	1.6	1.7	1.7	1.7
Inchworm	1.00	1.07	1.21	1.38	1.48	5.0	9.7	15.7	22.7	29.7
Bowtie	1.00	1.63	3.04	5.77	5.35	3.9	4.9	4.9	4.2	6.4
Samtools sort	1.00	1.64	2.95	5.09	5.01	4.7	6.0	6.1	5.8	8.3
Samtools view	1.00	1.44	2.11	2.86	2.78	4.9	7.1	8.9	10.7	15.6
Scaffold_iworm_contigs	1.00	0.99	0.99	0.99	1.00	1.0	2.2	4.0	6.5	9.2
GraphFromFasta	1.00	1.36	1.67	2.01	2.27	8.8	13.4	20.1	27.3	34.2
ReadsToTranscripts	1.00	2.31	4.08	5.97	9.42	39.6	35.6	36.9	41.3	36.9
Sort	1.00	1.13	1.05	0.81	1.00	0.4	0.7	1.5	3.1	3.6
Butterfly	1.00	1.40	2.33	4.52	7.00	28.0	41.3	45.7	38.5	35.1

with its runtime share.[1] The runtime share was additionally helpful to prioritize our optimization efforts. Next to the above discussed components, Table 3.1 reveals poor scaling in *GraphFromFasta* hidden in the Collectl graph, which results in over one third of runtime share. The in-depth analysis of the parallel behavior with Score-P and Vampir is detailed in Sect. 3.3.5.

3.3.2 Optimization of Samtools and Scaffold_iworm_contigs

Figure 3.1 shows a summation of the total CPU time used by each Trinity subcomponent. Thus, areas where only a very small number of CPU's were in use were of particular concern. One such area is evident from the 30 min mark to the 45 min mark. For approximately 10 % of the total runtime, two processes ran: *Scaffold_iworm_contigs* utilizing a single CPU and *Samtools view* which used almost no CPU, and no other processes are running at that time. The minimal CPU usage seemed to indicate a prime location for performance improvement.

Investigating this area further, we found the following: A tool that had run earlier in the process (*Bowtie*) had produced a Binary Sequence Alignment/Map (BAM) file. The information in the file was extracted to text using the standard bioinformatics tool suite *Samtools*, and the text was processed by *Scaffold_iworm_contigs*, a Perl script. Assuming that it would be more effective to parallelize the processing in C++, we ported the Perl script to a new code in that language. This, in turn, allowed us to take advantage of a *Samtools* C library which read the BAM file directly, and thus

[1] Since all components are called again during the *Butterfly* stage runtimes may deviate from the Collectl chart in Fig. 3.1.

removed the need to extract the BAM file to text. After testing these changes, we determined that the processing time required for this area of the application had fallen to 1 % or less of the total time, equaling a runtime reduction of about 10 min. Almost the entire time spent in this area of the application had been spent in converting the binary file to text and processing the text. It was then concluded that a further parallelization of the code was unnecessary.

3.3.3 Optimization of Sort

The CPU chart in Fig. 3.1 shows that *Sort* runs for about 6 min to sort a large text file containing the distribution of reads to components. While Trinity already tries to utilize the parallel version of *Sort*, the Collectl chart reveals that *Sort* runs only in serial mode. On further investigation we discovered that all the reviewed systems that run Trinity in production mode use *Sort* from the GNU coreutils 8.4 version. However, a parallel version of *Sort* was not introduced until GNU coreutils in the version 8.6 from October 2010. Consequently, we installed a current version of the GNU coreutils on the productive Trinity systems.

In addition, we discovered that the memory requirements of *Sort* increase logarithmically with the number of parallel threads. As a result, the memory limitation that can be passed to Trinity with the option `--max_memory` can critically limit the performance of the parallel *Sort*. For some data sets we recorded memory requirements larger than 200 GiB, which were available on the productive systems since Trinity requires large memory allocations in other stages, as well. Hence, we advise adapting Trinity's `--max_memory` option to the actual available memory to retrieve optimal performance in *Sort*.

Applying both of these sorting optimizations, the parallel version of *Sort* and a maximal memory limit, the runtime of *Sort* is decreased from over 6 min to 40 s.

3.3.4 Optimization of ReadsToTranscripts

Next to the more obvious optimization targets, *ReadsToTranscripts* revealed an overhead of frequent forking and joining of parallel regions. Within the main loop a predefined number of reads is loaded and than processed in a parallel region. The repeatedly forking and joining of the parallel region can be seen by the sharp drops of parallel CPU utilization in the Collectl chart (Fig. 3.1) and causes unnecessary overhead due to thread creation.

By increasing the default number of reads to be loaded from 10 to 50 million, the number of thread spawns was drastically reduced and the total runtime of *ReadsToTranscripts* was reduced by about 6 min. In addition, we identified an unnecessary entering of the parallel region if no additional reads were loaded, i.e., after all reads had been processed a final loop iteration works on zero input data. By leaving the loop iteration directly when no additional reads are available, we additionally reduced the runtime of *ReadsToTranscripts*. This change is particularly effective in the *Butterfly* stage where only a few reads are processed and the main loop is only processed once instead of twice. This resulted in a minor decrease of the runtime of *Butterfly* of about one minute.

3.3.5 Optimization of GraphFromFasta

Along with the above discussed components, Table 3.1 reveals poor scaling in *GraphFromFasta*. Since Collectl's CPU utilization (Fig. 3.1) did show that all 16 CPUs are utilized most of the time, it could be inferred that the poor scaling is the result of ineffective parallel patterns. To further investigate the issue we recorded the parallel behavior with the event-based trace monitor Score-P and analyzed it with Vampir.

Due to the massive data volumes and application slow down involved with tracing using automatic compiler instrumentation we recorded and analyzed the *GraphFromFasta* component using only a small test data set. Our initial assumption that the poor scaling originates from a load imbalance between the OpenMP threads proved to be wrong, since only a little time was spent in OpenMP critical sections or synchronization operations.[2]

To get further detail and reduce application slowdown, we recorded the application again without automatic compiler instrumentation but with manual instrumentation of the statements in the main loop. In addition, we recorded all OpenMP operations. Figure 3.2 shows the runtime behavior in comparison for one, two, four, eight and 16 threads in Vampir from top to bottom with white, red, yellow, green, and blue background, respectively. The left side depicts the active function over time on the horizontal axis and the different threads on the vertical axis. The accumulated exclusive runtime over all threads for each code region is represented in the function summary on the right side.

The analysis, represented in Fig. 3.2, revealed that the work load in the first part of *GraphFromFasta* increased nearly linearly with the number of OpenMP threads resulting in practically no parallel speed up with more than two threads. The cause for this was the frequent creation and destruction of string stream objects within an inner loop of the frequently called function *is_simple*. The string stream creation was internally locked by a mutex which resulted in excessive wait time, since all threads simultaneously created the string stream objects with a very high frequency. This can be seen by the increasing amount of time spent in the code region creating the string stream object from about 25 to 260 s.

By moving the string stream creation out of the inner loop and only clearing the string streams in the inner loop, we were able to avoid the serialization in this critical section. This resulted in a drastically increased parallel scaling and, therefore,

[2]Note: the load imbalance shown in Fig. 3.2 only occurs for the small test data set. For regular data sets the load is almost balanced.

Fig. 3.2 Resource utilization of original Trinity 2.0.6 version for a small test data set

a remarkable reduced runtime for the first part of *GraphFromFasta*. In addition to the better scaling, the serial runtime was reduced, as well; for the test data set, the serial runtime was reduced from 72 to 45 s. Figure 3.3 shows the improved scaling of the optimized version in a Vampir timeline for a 1 GiB test data. In this case the parallel speed up was increased to 8.9 instead of 2.3 with the unoptimized version. For the *S. Pombe* data set with 50 million base pairs the runtime of *GraphFromFasta* was reduced from 18:26 to 4:40 min.

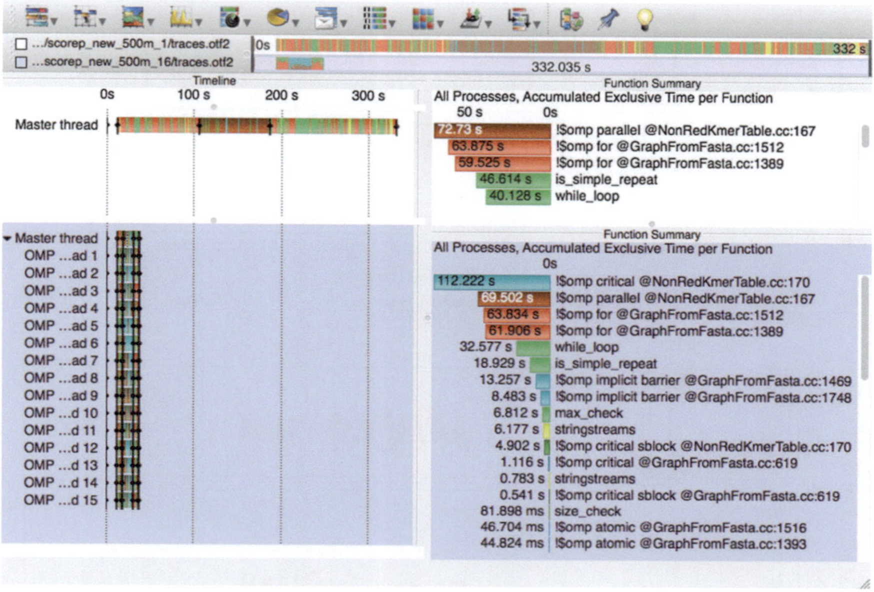

Fig. 3.3 Resource utilization of optimized Trinity version for medium size data set

3.3.6 *Optimization Results*

In total, the optimizations reduced the initial runtime of the 50 million base pair *S. Pombe* data set from 02:20 h by 31 min to 01:49 h, which equals a reduction of 22 %. Figure 3.4 shows the resource utilization of the original Trinity 2.0.6 version in comparison to the version including the above described optimizations. It highlights the runtime reduction in *Scaffold_iworm_contigs* due to the optimization in *Samtools*, in *GraphFromFasta*, *ReadsToTranscripts*, and *Sort*.

3.4 Tool Challenges and Restrictions

Although, performance analysis tools have been incredibly helpful in the above described optimization, their use was not without certain pitfalls and limitations. This section highlights some of the major issues we encountered: first, low sampling rates bias timing, second, busy-waiting is not detectable in a basic CPU utilization chart, third, tracing is unfeasible for nonstandard programming languages, fourth, tracing can result in massive data volumes and application slow down, and, finally, manual instrumentation was needed to narrow down the problem.

Collectl's five second interval biases timing of components. Running many of the tools required us to calculate a tradeoff: Was the amount of data that was collected

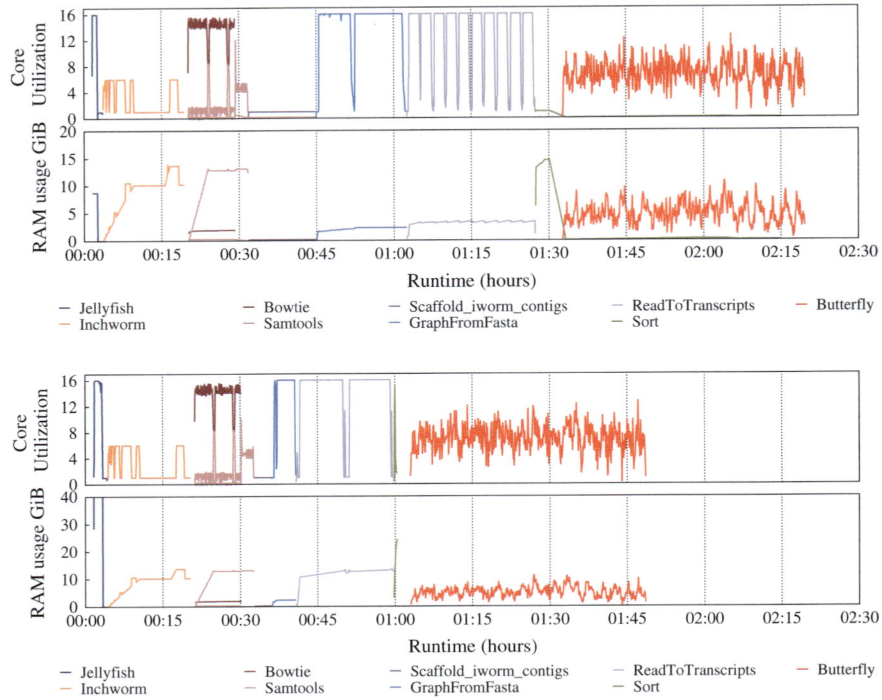

Fig. 3.4 Resource utilization of original Trinity 2.0.6 version (*top*) and optimized version (*bottom*)

sufficient to accurately determine performance bottlenecks, while at the same time being limited enough to find useful results? Collectl, for example, defaults to logging performance counters once per second. In order to minimize the impact on actual performance, the Collectl monitor that is built into Trinity records only once every five seconds. We chose to keep this default. However, certain embarrassingly parallel sections of Trinity may launch codes that complete in less than five seconds, and the time spent in these codes may not register with Collectl. As a result, our analysis of the total CPU usage of, for example, *Butterfly*, may underestimate the actual CPU usage.

Busy-waiting is not detectable in Collectl's CPU chart. A casual glance at the performance graph of Collectl is quite encouraging: several of the codes are displaying near-perfect usage of CPU. For example, the *ReadsToTranscripts* component, which runs from 40:00 to 60:00, shows very consistent usage of all 16 available Cores. An analysis with a different tool shows different information, however: *ReadsToTranscripts* is spending a lot of time waiting in OpenMP critical sections. Many implementations of OpenMP use busy-waiting in critical sections, and Collectl is unable to differentiate between this busy-waiting and actual CPU time.

Instrumentation is difficult for modules written in nonstandard programming languages. Trinity is a multi-language application: many sections are written in C++, while others are written in Java. A few are also written in Python, and the "glue" which holds the pipeline together is written in Perl. However, most HPC performance analyzers focus on standard languages such as C/C++ or Fortran. Other than with tools relying on system information like Collectl, it was not possible to record nonstandard language modules or even the entire pipeline. Therefore, an in-depth performance analysis with Score-P and Vampir could only be applied to a subset of components. In addition, it was necessary to break down the Trinity pipeline in a way allowing to monitor the components individually.

Tracing results in massive data collection and application slowdown. An in-depth analysis of individual components was additionally constrained by the massive data collection and application slow down due to the monitoring with Score-P using automatic compiler instrumentation. Both constrains appeared especially during the monitoring of the C++ components; a well-known issue for programming concepts frequently using small helper functions or get and set class methods [9]. As result, we had to turn of automatic compiler instrumentation and use smaller test data sets to limit trace sizes and application slow down.

Only manual instrumentation helped identifying the problem. Event tracing tools such as Score-P rely on functions as smallest recorded entities. While this is sufficient for many applications, in codes using a flat call hierarchy this can lead to too little detail. During the analysis of *GraphFromFasta*, with function instrumentation we were able to narrow down the issue to a function with 70 lines of code including two nested for loops, with only small clues as to where to look further. Only a manual instrumentation of individual statements revealed the thread contention during the creation of string streams highlighted in Fig. 3.2. However, manual instrumentation requires detailed knowledge of both the monitored application and the trace monitor.

3.5 Conclusion

This paper highlights our efforts in analyzing and optimizing Trinity, a well-established framework for the de novo reconstruction of transcriptomes from RNA-seq reads. With the help of the performance tools Collectl, Score-P, and Vampir we identified bottlenecks in the pipeline, with diverse causes including thread contention, suboptimal I/O, and streaming inefficiencies. We optimized the runtime behavior of the components *Samtools*, *Scaffold_iworm_contigs*, in *GraphFromFasta*, *ReadsTo-Transcripts*, and *Sort*. In total, the optimizations reduced the initial runtime of a 50 million base pair *S. Pombe* reference data set from 02:20 h by 31 min to 01:49 h, which equals a reduction of 22 %.

Acknowledgments We like to thank the Score-P team, in particular, Ronny Tschüter and Bert Wesarg for their friendly and prompt support.

References

1. Carrier, P., Long, B., Walsh, R., Dawson, J., Sosa, C.P., Haas, B., Tickle, T., William, T.: The impact of high-performance computing best practice applied to next-generation sequencing workflows. In: bioRxiv (2015). http://dx.doi.org/10.1101/017665
2. Collectl. http://collectl.sourceforge.net
3. Eschweiler, D., Wagner, M., Geimer, M., Knüpfer, A., Nagel, W.E., Wolf, F.: Open trace format 2: the next generation of scalable trace formats and support libraries. In: Applications, Tools and Techniques on the Road to Exascale Computing. Advances in Parallel Computing, vol. 22, pp. 481–490. a, a (2012)
4. Grabherr, M.G., Haas, B.J., Yassour, M., Levin, J.Z., Thompson, D.A., Amit, I., Adiconis, X., Fan, L., Raychowdhury, R., Zeng, Q., Chen, Z., Mauceli, E., Hacohen, N., Gnirke, A., Rhind, N., di Palma, F., Birren, B.W., Nusbaum, C., Lindblad-Toh, K., Friedman, N., Regev, A.: Full-length transcriptome assembly from RNA-Seq data without a reference genome. Nat. Biotechnol. **29**(7), 644–652 (2011)
5. Haas, B.J.: Trinity Release v2.0.6. http://github.com/trinityrnaseq/trinityrnaseq/releases/tag/v2.0.6
6. Henschel, R., Lieber, M., Wu, L.-S., Nista, P.M., Haas, B.J., LeDuc, R.D.: Trinity RNA-Seq assembler performance optimization. In: Proceedings of the 1st Conference of the Extreme Science and Engineering Discovery Environment: Bridging from the eXtreme to the campus and beyond (XSEDE'12), pp. 1–8 (2012)
7. Knüpfer, A., Brunst, H., Doleschal, J., Jurenz, M., Lieber, M., Mickler, H., Müller, M.S., Nagel, W.E.: The vampir performance analysis tool set. In: Tools for High Performance Computing, pp. 139–155 (2008)
8. Knüpfer, A., Rössel, C., an Mey, D., Biersdorff, S., Diethelm, K., Eschweiler, D., Geimer, M., Gerndt, M., Lorenz, D., Malony, A., Nagel, W.E., Oleynik, Y., Philippen, P., Saviankou, P., Schmidl, D., Shende, S., Tschüter, R., Wagner, M., Wesarg, B., Wolf, F.: Score-P: a joint performance measurement run-time infrastructure for periscope, Scalasca, TAU, and Vampir. In: Tools for High Performance Computing 2011, pp. 79–91 (2012)
9. Wagner, M., Doleschal, J., Knüpfer, A., Nagel, W.E.: Selective runtime monitoring: non-intrusive elimination of high-frequency functions. In: Proceedings of the International Conference on High Performance Computing & Simulation (HPCS), pp. 295–302 (2014)

Chapter 4
Power Management and Event Verification in PAPI

Heike Jagode, Asim YarKhan, Anthony Danalis and Jack Dongarra

Abstract For more than a decade, the PAPI performance monitoring library has helped to implement the familiar maxim attributed to Lord Kelvin: "If you cannot measure it, you cannot improve it." Widely deployed and widely used, PAPI provides a generic, portable interface for the hardware performance counters available on all modern CPUs and some other components of interest that are scattered across the chip and system. Recent and radical changes in processor and system design—systems that combine multicore CPUs and accelerators, shared and distributed memory, PCI-express and other interconnects—as well as the emergence of power efficiency as a primary design constraint, and reduced data movement as a primary programming goal, pose new challenges and bring new opportunities to PAPI. We discuss new developments of PAPI that allow for multiple sources of performance data to be measured simultaneously via a common software interface. Specifically, a new PAPI component that controls power is discussed. We explore the challenges of shared hardware counters that include system-wide measurements in existing multicore architectures. We conclude with an exploration of future directions for the PAPI interface.

4.1 Introduction

Most of the major tools that high-performance computing (HPC) application developers use to conduct low-level performance analysis and tuning of their applications typically rely on hardware performance counters to monitor hardware-related activities. The kind of counters that are available is highly hardware dependent; even across the CPUs of a single vendor, each CPU generation has its own implementation. The PAPI performance-monitoring library provides a clear, portable interface to the hardware performance counters available on all modern CPUs, as well as some GPUs, networks, and I/O systems [1, 8, 9, 13]. Additionally, PAPI supports trans-

H. Jagode (✉) · A. YarKhan · A. Danalis · J. Dongarra
Innovative Computing Laboratory, University of Tennessee, Knoxville,
Knoxville, TN 37996, USA
e-mail: jagode@icl.utk.edu

A. Knüpfer et al. (eds.), *Tools for High Performance Computing 2015*,
DOI 10.1007/978-3-319-39589-0_4

parent power monitoring capabilities for various platforms, including Intel Xeon Phi and Blue Gene/Q [10]—enabling PAPI users to monitor power in addition to traditional hardware performance counter data, without modifying their applications or learning a new set of library and instrumentation primitives.

With the increase in scale, complexity, and heterogeneity of modern extreme scale systems, the design of future HPC machines will be driven by energy efficiency constraints. Hence, the ability to control power and energy consumption has become one of the critical features for future development. To allow the HPC community to not only use PAPI to monitor but also manage power consumption, PAPI has been extended with a component supporting power writing capabilities. This paper provides detailed information describing this new component and explores its usefulness with different case studies.

At the same time, the rapid changes and increased complexity we have witnessed in processor and system design—with systems that combine multicore CPUs and accelerators, shared and distributed memory, PCI-express and other interconnects—require a continuous series of updates and enhancements to PAPI with richer and more capable methods that are needed to accommodate these new innovations. Extending PAPI to monitor performance-critical resources that are shared by the cores of multicore and hybrid processors, including on-chip communication networks, memory hierarchy, I/O interfaces, and power management logic, will enable tuning for more efficient use of these resources. Failure to manage the usage and, more importantly, contention for these "inter-core" resources has already become a major drag on overall application performance. We discuss one of PAPI's new features: the Counter Inspection Toolkit (CIT), which is designed to improve the understanding of these inter-core events. Specifically, the CIT integrates micro-benchmarking based methods to gain a better handle on off-core/un-core/NorthBridge related events. We aim to define and verify accurate mappings between particular high-level concepts of performance metrics and underlying low-level hardware events. This extension of PAPI engages novel expertise in low-level and kernel-benchmarks for the explicit purpose of collecting meaningful performance data of shared hardware resources.

In this paper we outline a new PAPI component that supports power and energy controlling through the Intel RAPL interface. A detailed description of the power writing capabilities is provided in addition to case studies that validate the usage of this component. Further, we briefly describe the objectives of the PAPI Counter Inspection Toolkit, and then focus on the micro-kernels that will be used to measure and correlate different native events.

4.2 Power Management Using the Intel RAPL Interface

As processors have grown more complex, power consumption has also increased. This increase in power consumption not only results in increased power costs, but also increased costs for managing the waste heat that is generated by the components. This power and cooling situation is exacerbated by the sizes of data centers and supercomputers. For some time now, interest in managing this power consumption has been increasing, and hardware manufacturers have been introducing low level controls that allow users to make trade-offs between power and computational performance.

Starting with the Sandy Bridge architecture, Intel has included a RAPL (Running Average Power Limit) interface for accessing and managing power features on the processor [5]. This interface exposes features designed for thermal management to a more general user-space. Much of the RAPL interface is accessed by reading and writing information to MSRs (Model Specific Registers).

Current directions for PAPI development include providing applications the ability to trade-off power for performance. To this end, we are experimenting with a PAPI component that will include control aspects, i.e., this component has an active interface that **writes** values to the RAPL/MSR interface. This is a significant change from all prior PAPI components which have had an entirely passive measurement interface to **read** information and events.

This new RAPL/MSR component is expected to, eventually, actively handle interrupts (thermal limits) and change system states (set per-core clock gating, set per-node power cap, enable/disable turbo mode). The following section discusses our early prototype version of the RAPL/MSR component, which has not yet been released.

Providing users with unrestricted access to write data to MSRs (Model Specific Registers) can have many significant performance and security implications. In order to encourage system administrators to give wider access to the MSRs on a machine, LLNL has released a Linux kernel module (`msr_safe`) which provides safer, white-listed access to the MSRs that can be tuned by the site administrator [11]. Lawrence Livermore National Laboratory scientist Barry Rountree has released a library (libmsr) to provide a simple, safe, consistent interface to several of the model-specific registers (MSRs) in Intel processors via the `msr_safe` kernel module [15].

PAPI has created a component that can provide read and write access to the information and controls exposed via libmsr. A scientist can use the well-known, standardized PAPI API to read the state of power consumption on a CPU socket and the current performance of the code, make determinations about the desired CPU performance, and adjust and cap the power consumption as desired. RAPL allows users to set power limits over two specific time windows—meaning, one can have local power spikes, while still keeping the power low over a larger time window. One example of a situation where this might be of interest is when a scientist is aware that the computation requirements will decrease due to communication (I/O bound) and that the overall execution time will not suffer if the CPU power is capped temporarily.

Figure 4.1 shows a simple example using PAPI to measure and adjust the power consumption of an iterative program where each iteration does one consistent unit of work. Initially the power is at a default high level so the performance of the unit task is high. Then we attempt to cap the power at a very low level, below the minimum allowed for the CPU. The power drops to the allowed minimum and the computation time for a unit of work increases. At each 10th iteration the power cap is increased, and the time taken for a unit of work decreases. Finally, we attempt to increase the

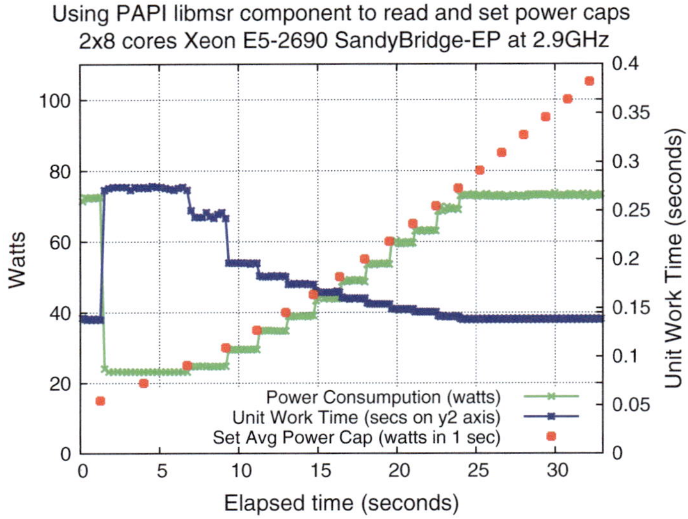

Fig. 4.1 Power controlling with PAPI example

power cap higher than the allowed maximum for the machine. At this point, time taken for a unit of work stays consistent. This simple demonstration shows PAPI's new ability to write information to a counter as well as reading data from a counter.

4.2.1 Case Study: LU Factorization

Here we consider a different usage scenario where controlling power may be appropriate. Many applications can be decomposed as dataflow DAGs (Directed Acyclic Graph), with data dependencies between the various tasks that compose the application. This approach to structuring applications as dataflow DAGs is increasing in popularity with the advent of heterogeneous hardware platforms with large numbers of computation resources, since programming and scheduling on such platforms is a challenge. Dataflow DAGs can be efficiently managed by a runtime and tend to have good load balancing and efficiency characteristics.

If we have an application that is instantiated as a dataflow DAG, then there is often a critical path of tasks that determines the overall execution time of the application. An opportunity for saving power could exist if we schedule all the tasks in the critical path on fast resources, and then schedule the remaining tasks on sockets with decreased power consumption. Under certain circumstances, it can be possible to execute the application without any loss of overall execution time while saving power. However, in most cases there will be some increase in overall execution time.

As an example of a dataflow DAG computation, we consider the tile-based implementation of the LU factorization of a matrix as described in pseudocode in Fig. 4.2a.

(a)

```
GETRF( A ) {
  for (k = 0; k < M; k++) {
     // panel factorization
     GETRF( A(k:M-1,k) );
     for (n = k+1; n < N; n++) {
        // row interchanges to the right
        LASWP ( A(k:M-1,n) );
        // triangular solve at the top
        TRSM ( A(k,n) );
        for (m = k+1; m < M; m++) {
           // matrix multiply to the right
           GEMM( A(m,k), A(k,n), A(m,n) );
        }
     }
  }
  for (k = 1; k < M; k++) {
     // row interchanges to the left
     for (n = 0; n < k-1; n++) {
        LASWP( A(k:M-1,n) );
     }
  }
}
```

(b)

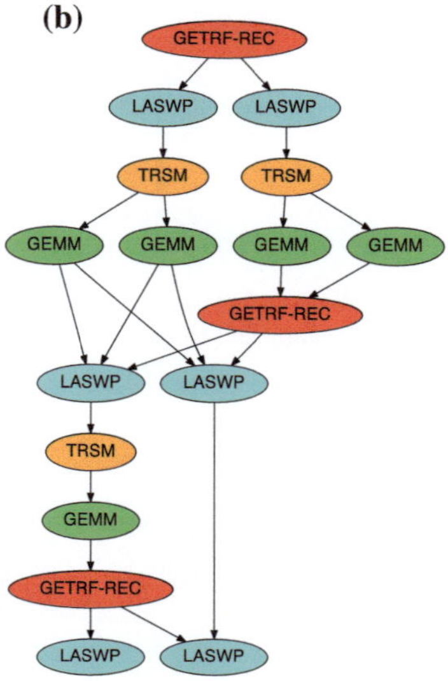

Fig. 4.2 On the *left* we have an LU factorization algorithm that is expressed as tasks (i.e., GETRF, LASWP, TRSM, GEMM) acting on data items (i.e., A(i, j)). When these tasks are executed, the execution can be viewed as a dataflow DAG, where the vertices are the tasks and the edges are data dependencies between them. **a** LU Factorization. **b** LU Factorization DAG

The details of this implementation are beyond the scope of this paper, for more information on the LU DAG implementation, the reader is referred to [2, 7]. In a nutshell, each function corresponds to a task that works on a unit of data—either a tile or a row or an entire panel of data. For example, the red GETRF tasks are large complex tasks that work on an entire panel of data, while the other tasks work on smaller chunks of data and execute much faster.

The DAG of the execution is shown in Fig. 4.2b which reveals that the red GETRF tasks are in the critical path of the graph. Since the GETRF tasks operate on an entire panel of data and are relatively inefficient, they take longer to execute than the other tasks. Our goal is to take this knowledge into consideration by scheduling these expensive GETRF tasks on faster cores because we want them to execute as fast as possible. Other tasks that are not on the critical path are allowed to execute on sockets and cores where the power usage is decreased and where they may run more slowly. Ultimately, this approach will enable us to save energy at a minimal cost to the overall execution time as determined by the critical path.

We conducted this experiment on a 2.90 GHz Intel Xeon Sandy Bridge E5-2690 system, and here we show some small scale results using two sockets (four cores per socket). In Fig. 4.3 we have a small trace from the execution of the dataflow DAG for LU factorization. We can see that the red GETRF is taking a large amount of time as expected. Since this slow task is in the critical path of the execution, there are many occasions where the other CPUs complete all other available tasks and are in an idle state waiting for the GETRF to complete.

Fig. 4.3 Both socket 0 and socket 1 running at full power. Note that the panel factorization GETRF task (*red*) is long and on the critical path, so there is white space where no tasks are available to run on socket 1

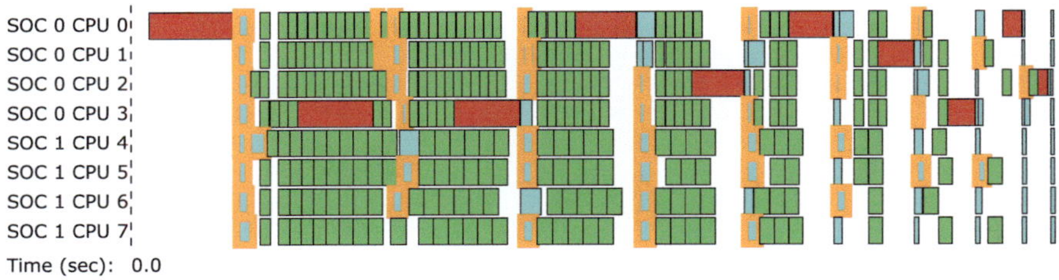

Fig. 4.4 Slow down socket 1 using RAPL and lock critical path GETRF tasks to socket 0. The GEMM tasks (*green*) take longer, filling out the white space on socket 1. This occurs without any overall loss in time for the full execution

This leads us to our opportunity to save power without affecting the overall computation time. We use the runtime environment to restrict the GETRF task to run on socket 0. We use the PAPI libmsr component to write to the RAPL MSRs in order to limit and decrease the power consumption at socket 1. This causes the tasks assigned to socket 1 to take longer, so socket 1 has a higher level of occupation (at a lower power) and does not have as much idle time. In Fig. 4.4 we observe that the individual green GEMM tasks on socket 1 take longer to execute than on socket 0, absorbing the idle time that was wasted in the previous trace. In this small example, the overall computation time was unaffected by slowly reducing the power consumption on socket 1.

We now extend to a larger scale experiment, where we run the tile LU factorization on a matrix that is of size $N = 17920$, consisting of 80×80 tiles of 224×224 double precision numbers. Once again this experiment is run on a 2.90 GHz Intel Xeon SandyBridge E5-2690 system using two sockets (four cores per socket). For this larger problem, the low power execution achieves a small decrease in total power consumed (4001 J) when compared to the high power execution (4136 J). However, the overall time is increased during the low power execution. This is because for the larger matrix size there are so many GEMM tasks generated for each slow POTRF task, that there is minimal idle time for the low power execution to absorb. For other types of dataflow DAG computation, where there is sufficient idle time, this technique may be able to save power while maintaining the overall execution time (Fig. 4.5).

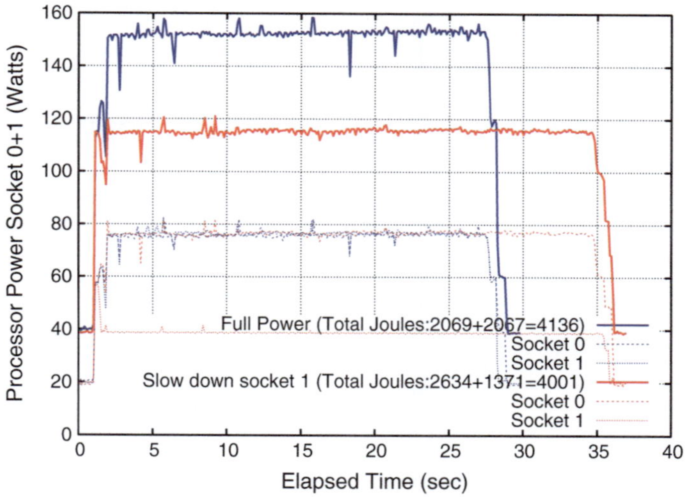

Fig. 4.5 Tiled LU (N=17920=224 × 80) using a SandyBridge EP (2 sockets, 4 cores/socket)

4.3 Counter Inspection Toolkit

In modern architectures, native events count behaviors that relate to the specifics of a particular architecture, but do not necessarily map to higher level concepts of performance. For example, when a memory region is accessed by an application, assessing the level of cache reuse is important when studying the performance of that application. One would expect that this can be done using a native event that counts cache misses, such as the events `LLC_MISSES` and `LLC-LOAD-MISSES`—which measure the last level cache (L3) misses.

However, our experiments show that the numbers reported by PAPI do not always match expectations. Specifically, LLC measurements do not match the expected behavior of the micro-benchmarks in the BlackjackBench [3] suite, which we developed in our previous work. One of the key goals of BlackjackBench was to characterize the cache hierarchy, so special attention was given to stressing different levels of the cache hierarchy. The way this is achieved is through the use of pointer chaining.

Figure 4.6 shows an abstracted code snippet and a schematic outline of pointer chaining. The key idea is that the benchmark is split into a setup phase and a measuring phase. During the setup phase an array is allocated and each element is made to point to another element (i.e., every element stores the address of another element of the array). This creates a chain between the elements of the array. Since this is done in a setup phase—during which neither time, nor hardware events are being measured—the process can be arbitrarily expensive. As a result, every time we need to select the next element of the chain we utilize the POSIX function `random()`, which has a very large period (approximately $16 * ((2^{31}) - 1)$) and good (albeit not cryptographically strong) randomness properties. In the context of our benchmark, good randomness

```
SETUP( ) {
   p = (uintptr_t **) &array[0];
   for (i = random( ) ) {
      next = &array[i];
      *p = next;
      p = (uintptr_t **) next;
   }
}

MEASURE( ) {
   start_measurement();
   for (...) {
      p = (uintptr_t **) *p;
   }
   stop_measurement();
}
```

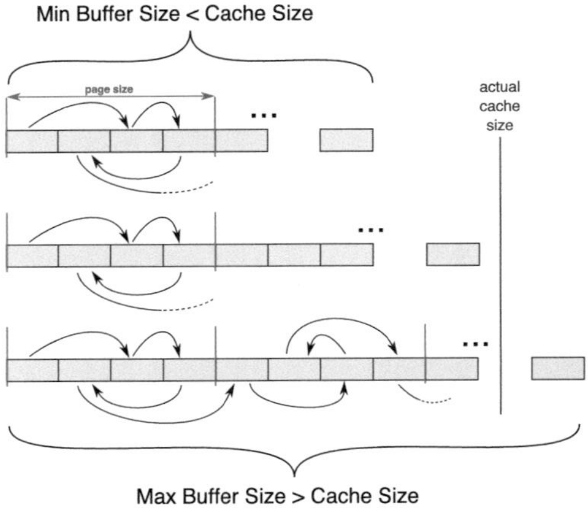

Fig. 4.6 Random pointer chaining

properties means that when we traverse the chain during the measurement phase the memory access sequence will consist of jumps that do not obey any regular patterns that the hardware prefetcher could guess.

Clearly, the actual setup code is more complex than shown in the pseudocode of the figure in order to ensure that the next element that we choose is neither an element that has already been used, nor the current element, and that each pointer is aligned properly so that we only access one element per cache line. Also, as shown in the figure, we populate the buffer one page at a time, in order to stress the cache without stressing the TLB. This is achieved by using a modulo operation (`%`) to keep only the lower bits of the numbers returned by `random()`. Finally, when we reach the last element, we make it point back to the first, so that we can traverse the chain multiple times without explicitly starting and stopping the traversal.

After the setup phase has completed, the measuring phase starts by initiating the desired counters. Then we traverse the pointer chain using a simple loop that dereferences each element to find the next, and when the traversal is over we read the values of the counters. As we discussed earlier, the non-trivial randomness of the chosen setup method ensures that there will be few (if any) regular patterns in the memory traversal, or at least not enough to affect the overall execution time due to hardware prefetching. Furthermore, compilers are incapable of optimizing the code that traverses the chain, since the location of each memory access depends on program data (i.e., the next address is always the value read in the previous array access). As a result, all elements will be explicitly accessed by the code and the probability of any of them having been prefetched by the hardware is low.

In BlackjackBench, we used this technique to setup our arrays and then measured the time it took to traverse such an array as a function of the array size. As shown in Fig. 4.7 the access latency per element jumps every time the array size exceeds the size of a different level of cache, and stays constant while the array fits in a given level of the cache hierarchy. The accuracy of these results make us fairly confident

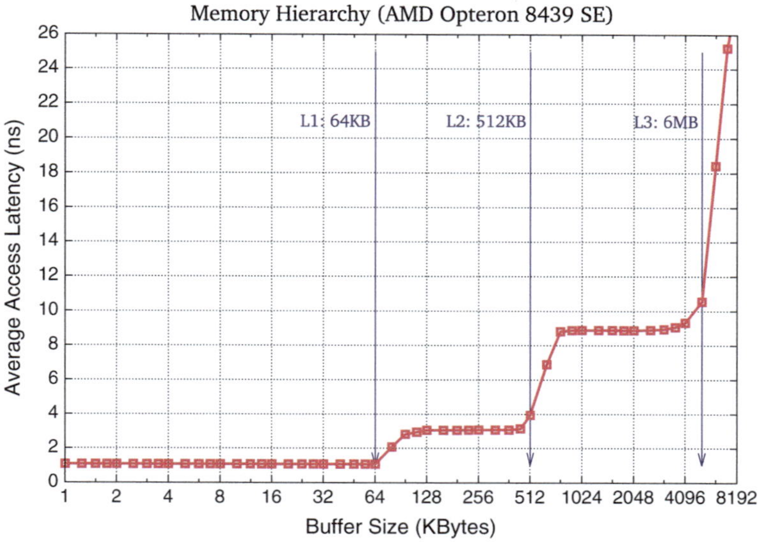

Fig. 4.7 Benchmark timing

in the behavior of the benchmark. However, this same benchmark, when used in conjunction with PAPI native events to measure LLC misses, gives us inconsistent results. Depending on the exact native event used and the architecture on which the experiment takes place, the resulting measurement can be more than expected, less than expected, or even zero!

Given this mismatch, there is a need for a way to *validate* native events, or in other words, assess whether the value a native event measures matches what a human developer thinks it is measuring. Furthermore, there is a need to *define* high level "predefined" events that combine the values of native events in order to provide measurements that match developer intuition. The need for validation and assisted definition of high level events is becoming increasingly urgent in the context of inter-core resource counters, as understanding and utilizing those is particularly challenging.

The Counter Inspection Toolkit, which we are developing, will provide kernels that perform well-defined operations and will use them to measure native events. Subsequently, automatic analyses will attempt to correlate different native events, or combinations of native events to the high level operations. We expect the outcome of this research to be threefold. First, it will increase the portability of PAPI into new hardware with new native events. Second, it will assist PAPI developers in combining native events to define predefined events. Finally, it will provide customization by enabling the user of PAPI (whether that is a human developer of an application, or an additional layer of performance tools) to define custom combinations of events to fit whatever parameters interest a particular user.

4.4 Related Work

Although PAPI has been widely utilized by HPC users for many years, drawing on its strength as a cross-platform and cross-architecture API, there are other tools for gathering performance information like hardware counter data, profiling and tracing data, and MPI library state data.

The *perf tool* [12] makes use of the *perf_event* API which is part of the Linux kernel. Although perf_event attempts to provide a generic interface for Linux platforms, it is still very low-level and the information returned requires considerable interpretation to be useful to tool developers or end users.

Processor vendors supply tools for reading performance counter results. This includes *Intel VTune* [16], *Intel VTune Amplifier*, *Intel PTU* [6], and *AMD's CodeAnalyst* [4]. These program the CPU registers directly, avoiding the Linux kernel. Since the counter state is not saved on the context switch, only system-wide sampling is available, and there is also no API for accessing the results.

The *likwid* lightweight performance tools project [14] allows accessing performance counters by bypassing the Linux kernel and directly accessing hardware. This can have low overhead but can conflict with concurrent use of other tools accessing the counters. It can also expose security issues, as it requires elevated privileges to access the hardware registers and this can lead to crashes or system compromises. likwid provides access to traditional performance counters and also RAPL energy readings. Unlike PAPI, likwid is not cross-platform, only x86 processors are supported under Linux, and only system-wide measurements are available (counters are not saved on context-switch). Currently there is no API for accessing values gathered with likwid; a separate tool gathers the results and stores them in a file for later analysis.

4.5 Conclusion and Future Work

With larger and more complex high performance systems on the horizon, energy efficiency has become one of the critical constraints. To allow the HPC community to "control" power, in addition to the traditional hardware performance counter "monitoring" approach, PAPI has been extended with a component that supports power writing capabilities through the Intel RAPL interface. Whether PAPI is applied as a stand-alone tool or as a middleware by third-party performance analysis tools, the new PAPI component for power controlling can be used without the need for application developers to modify their applications or learn new library primitives.

Furthermore, we introduced PAPI's new Counter Inspection Toolkit, which will be fully integrated for future PAPI releases. It establishes methods to automatically determine which hardware event combinations map to particular high level concepts of performance.

Acknowledgments We thank the anonymous reviewers for their improvement suggestions.

This material is based upon work supported in part by the DOE Office of Science, Advanced Scientific Computing Research, under award No. DE-SC0006733 "SUPER—Institute for Sustained Performance, Energy and Resilience," and by the National Science Foundation under award No. 1450429 "PAPI-EX."

References

1. Browne, S., Dongarra, J., Garner, N., Ho, G., Mucci, P.: A Portable Programming Interface for Performance Evaluation on Modern Processors. Int. J. High Perform. Comput. Appl. **14**(3), 189–204 (2000)
2. Buttari, A., Langou, J., Kurzak, J., Dongarra, J.: A class of parallel tiled linear algebra algorithms for multicore architectures. Parallel Comput. **35**(1), 38–53 (2009)
3. Danalis, A., Luszczek, P., Marin, G., Vetter, J.S., Dongarra, J.: BlackjackBench: Portable hardware characterization with automated results' analysis. Comput. J. **57**(7), 1002–1016 (2013)
4. Drongowski, P.: An introduction to analysis and optimization with AMD CodeAnalyst™Performance Analyzer. Advanced Micro Devices, Inc. (2008)
5. Intel, I.: Intel 64 and IA-32 Architectures Software Developer's Manual - Systems Programming Guide, vol. 3, chap. 14 (2015)
6. Intel™ Performance Tuning Utility. http://software.intel.com/en-us/articles/intel-performance-tuning-utility/
7. Kurzak, J., Luszczek, P., YarKhan, A., Faverge, M., Langou, J., Bouwmeester, H., Dongarra, J.: Multithreading in the PLASMA Library. Handbook of Multi and Many-Core Processing: Architecture, Algorithms, Programming, and Applications. Computer and Information Science Series. Chapman and Hall/CRC, Boca Raton (2013)
8. Malony, A.D., Biersdorff, S., Shende, S., Jagode, H., Tomov, S., Juckeland, G., Dietrich, R., Poole, D., Lamb, C.: Parallel performance measurement of heterogeneous parallel systems with gpus. In: Proceedings of the 2011 International Conference on Parallel Processing, ICPP '11, pp. 176–185. IEEE Computer Society, Washington, DC, USA (2011)
9. McCraw, H., Terpstra, D., Dongarra, J., Davis, K., R., M.: Beyond the CPU: Hardware Performance Counter Monitoring on Blue Gene/Q. In: Proceedings of the International Supercomputing Conference 2013, ISC'13, pp. 213–225. Springer, Heidelberg, June (2013)
10. McCraw, H., Ralph, J., Danalis, A., Dongarra, J.: Power Monitoring with PAPI for Extreme Scale Architectures and Dataflow-based Programming Models, pp. 385–391 (2014)
11. McFadden, M., Shoga, K., Rountree, B.: Msr-safe (2015). https://github.com/scalability-llnl/msr-safe
12. Molnar, I.: perf: Linux profiling with performance counters (2009). https://perf.wiki.kernel.org/
13. Terpstra, D., Jagode, H., You, H., Dongarra, J.: Collecting performance data with PAPI-C. Tools for High Performance Computing 2009, pp. 157–173 (2009)
14. Treibig, J., Hager, G., Wellein, G.: LIKWID: A lightweight performance-oriented tool suite for x86 multicore environments. In: Proceedings of the First International Workshop on Parallel Software Tools and Tool Infrastructures September (2010)
15. Walker, S., Shoga, K., Rountree, B., Morita, L.: Libmsr (2015). https://github.com/scalability-llnl/libmsr
16. Wolf, J.: Programming Methods for the Pentium™ III Processor's Streaming SIMD Extensions Using the VTune™Performance Enhancement Environment. Intel Corporation (1999)

Chapter 5
Gleaming the Cube: Online Performance Analysis and Visualization Using MALP

Jean-Baptiste Besnard, Allen D. Malony, Sameer Shende, Marc Pérache and Julien Jaeger

Abstract Multi-Application onLine Profiling (MALP) is a performance tool which has been developed as an alternative to the trace-based approach for fine-grained event collection. Any performance and analysis measurement system must address the problem of data management and projection to meaningful forms. Our concept of a *valorization chain* is introduced to capture this fundamental principle. MALP is a dramatic departure from performance tool dogma in that is advocates for an online valorization architecture that integrates data producers with transformers, consumers, and visualizers, all operating in concert and simultaneously. MALP provides a powerful, dynamic framework for performance processing, as is demonstrated in unique performance analysis and application dashboard examples. Our experience with MALP has identified opportunities for data-query in MPI context, and more generally, creating a "constellation of services" that allow parallel processes and tools to collaborate through a common mediation layer.

5.1 Introduction

Scalable performance measurement on modern supercomputing systems inevitably becomes a problem of scalable data management. Whether parallel profiles or traces are collected, large, distributed performance data are a fundamental part of an application's execution. If the performance data is accessed during execution (we use the term "performance monitoring" in this case), the data access and transformation (i.e., analytics) becomes an important concern. Ultimately, this becomes a question of how performance information is eventually processed by a human, requiring a *projection* to an intelligible state that acknowledges the end-user's cognitive

J.-B. Besnard (✉)
ParaTools SAS, Bruyeres-le-chatel, France
e-mail: jbbesnard@paratools.fr

A.D. Malony · S. Shende
ParaTools Inc., Eugene, USA

M. Pérache · J. Jaeger
CEA, DAM, DIF, 91297 Arpajon, France

A. Knüpfer et al. (eds.), *Tools for High Performance Computing 2015*,
DOI 10.1007/978-3-319-39589-0_5

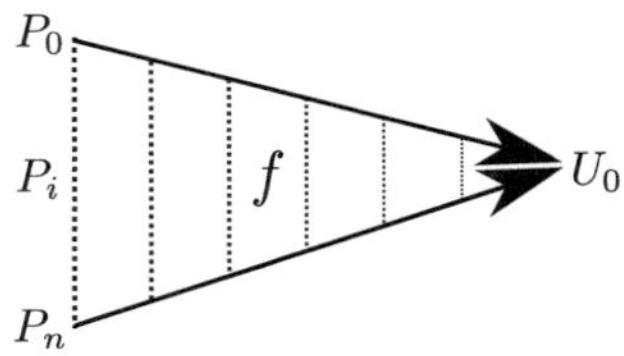

Fig. 5.1 Illustration of a measurement chain from processes ($P_{0\rightarrow n}$) to a single end-user (U_0)

capabilities. This defines a parallel measurement (processing) chain with the purpose to allow the exploration of a distributed performance state, for example, to understand it (debugging), to derive potential inefficiencies (performance) or to pinpoint misuses of standardized interfaces (validation). In this section we propose to discuss the trade-offs guiding the design of such measurement chains first from a global point of view and then by going through alternative approaches developed by tools.

5.1.1 From Performance Data Management to Performance Valorization

In order to define more formally the role of such measurement chain for HPC tools, consider Fig. 5.1. On the left we identify several processes $P_{0\rightarrow n}$ which are *spatially* distinct, being scattered in various nodes within the parallel machine. On the right, we modeled the end-user who desires the transformed performance information in more meaningful forms. Thus, the data have to be *projected* to global metrics which are easier to interpret, a process modeled by function f. This function is not necessarily statically defined and could be taking user-defined parameter. Moreover, this function is by definition performing data analytics associated with a computational cost. Dealing with the *projection* it can be done on any dimension defining a parallel computation: spatial (per process/thread/node), temporal, per procedures, per file, per programming model, and so on. However, in all cases it has to be doing some form of transformation (reduction, aggregation, analysis, ...) in order to provide actionable, semantic information to the end-user.

Starting from this simple model we can see that the measurement chain is closely associated with a *valorization*[1] function f which role is to transpose numerous events to an unified view interpretable by an human. It is interesting to note that this function also plays role in instrumentation chain's scalability by defining what are the valuable informations. Indeed, by definition a parallel computation will generate a large number of instrumented events from several processes, leading to a data-management problem. In the mean time, we argued that the user cannot take advantage of individual events and therefore that a tool was needed to explore these data either working on projections or subsets. It is then possible to envision projections valorizing data

[1] *Valorize* means to give or ascribe value or validity to something.

before display, for example, by reducing spatially during measurement—overcoming in the meantime data-management problematics.

It is this relationship between performance data and how and when that data are processed which gave rise to the landscape of performance tool approaches and infrastructures we see today.

5.1.2 In-Place Performance Data Processing

When processing data in-place, events of interest are projected at the moment they are instrumented. It has the advantage of processing data locally, avoiding to generate any event-related data movement (all can be done on the stack). The counterpart of this locality is the projection function cost which directly affects the application. As a consequence, this projection cost has to remain limited in terms of both computational power and memory footprint as it shares resources with the target. In such conditions, it is then generally prohibitive to maintain shared state. The in-place approach is among the most scalable, overcoming data-management problematic by performing an immediate projection. The *mpiP* [24] performance tool uses in-place processing to precisely track MPI time. It first performs a statistical temporal reduction before aggregating results spatially when application ends. The *HPCToolkit* [1] performance tool uses in-place sampling and statistical profiling to project execution time, counters, and other events on threads of execution before aggregating profiles spatially when application ends.

5.1.3 Trace-Based Performance Data Processing

The trace-based approach has been retained by several tools. It consists in storing all the events in a file-system trace in order to have them processed in a post-mortem fashion. This has several advantages, first the processing is completely decoupled from the instrumented (measured) program and can therefore have an arbitrary cost without impacting measurement quality. Moreover, as file-system is a non-volatile storage the trace can be analyzed several time at various granularities. Nonetheless, performance traces also bring several problems. First, the file-system is known to be subject to contention as a shared resource, its scalable usage requiring the use of a parallel IO library such as for example SionLib [10] used by Score-P [17]. Similarly, event verbosity can lead to impractical trace sizes. In the meantime as parallel IOs can be relatively expensive, instrumenting events with small durations can lead to prohibitive overheads.[2] A wide range of trace formats were developed, Pajé [20] with a focus on genericity, SLOG [6] which was designed for temporal trace visualization in Jumpshot [26], and the Open Trace Format 2 (OTF2) [9] (which replaced OTF1

[2]Dilation is a function of the duration ratio between events of interest and instrumentation cost.

[15] with several performance improvements). Using this approach, the Vampir [16] tool allows the interactive exploration of parallel application's temporal behavior, up to single event granularity.

5.1.4 OnLine Performance Data Processing

This last approach can be seen as a combination of the previous two. It consists in coupling instrumented processes with processes dedicated to the analysis at runtime. This removes the need for the data-multiplexing required by parallel IO libraries, directly using network coupling. In complement, having dedicated computing resources in the analysis allows more complex event processing, possibly with a distributed state. Nonetheless, as analysis processes are limited in memory, events still have to be either processed fast-enough or sufficiently small, limitation which can be mitigated by increasing processing resources. Such an approach has been used at scale with, for example, the MrNET Tree-Based Overlay Network (TBON) [21], performing a spatial reduction of stack traces in the STAT debugger [2]. This kind of reduction tree is also used to characterize applications in Periscope [3] or to derive a global state in the Generic Tools Infrastructure (GTI) [13] used by MUST [14] for validation. Similarly, commercial debuggers such as DDT and TotalView rely on a tree-based overlay network to control and monitor distributed processes. In MALP [5], we use this approach for fine-grained online trace analysis.

5.1.5 Summary

Table 5.1 summarizes the aforementioned performance processing approaches. While the trace-based approach traditionally decouples the analysis from the application's execution and allows replay, it still has scalability problems when facing large trace volumes that can force online actions (e.g., transferring in-memory trace buffers to disk when they become full).

Table 5.1 Performance coupling-modes characteristics overview

Approach	Bandwidth usage	Overhead decoupling	Analysis type	Replay
Trace-based	File-system	Total (post-mortem)	Arbitrary analysis	Yes
In-place	None (Stack)	None	Lightweight	No
Online	Network	Partial (pipeline)	Bandwidth dependent	No

Dealing with the Online approach, the notion of allocating extra resources in the online approach partially decouples the analysis from the instrumentation (measurement), thereby overcoming file-system limitations thanks to direct network coupling. Such a method supposes that performance events can be processed fast-enough so as not to impact the instrumented application, therefore putting an upper bound on analysis' complexity.

Clearly, nothing prevents these approaches to be combined to mitigate their limitations. For example, when we think of an IO proxy, one would use the online coupling paradigm to perform trace-based storage—delegating the IO overhead to remote processes furnishing their local memory as caches. Performance tools also mix these approaches. For example, Scalasca [12] reduces call-stacks locally and then unifies event identifiers during analysis to derive performance metrics, eventually storing an unified report for post-mortem processing—covering the three approaches we presented. Similarly, the Tuning and Analysis Utility (TAU) [23] has continually explored different approaches, proposing profiles, traces and even online results through MrNET [19]. TAU and Scalasca eventually contributed to the Score-P [17] unified infrastructure which also is able to collect call-stack profiles while generating traces, mixing in-place and post-mortem approaches.

Looking at all these state of the art tools deploying various data-management approaches, one can see the trade-offs implied by supercomputers where a global state is out of reach, context in which tools need to derive models and metrics more than raw data. Naturally, traces describe a parallel execution with the highest level of details thanks to re-playable fine-grained event. However, file-system limitations coupled with the combinatorial complexity of millions of computation units communicating advocate for a *multi-scale* profiling approach. Therefore, starting from a global overview generated through scalable spatial reductions such as TBONs or temporal/functional reduction inside local accumulators, one should be able to change the *focus of attention* to query local data—possibly more verbose as less spatially diffuse. This model supposes that performance data ranging from the local to the parallel state can be queried in a selective manner, somehow reversing the instrumentation chain. In other words, if a trace will always be the most verbose manner to explore a problem either spatially or temporally located, we think that at larger scales in-situ/online analytics and visualization can provide sufficient insights to heuristically extract valuable subsets from the unreachable parallel state.

The following section presents our research on the Multi-Application OnLine Profiling or MALP framework Its online nature was at first motivated by file-system's limitations, requiring the use of a secondary set of processes to perform the reduction while the application was running in order to avoid storing large volumes of redundant data. This allowed us to generate profiling reports without storing an intermediate trace as detailed in Sect. 5.3. Then, as our instrumentation chain was collocated with the application we started investigating in-situ approaches, trying to "see" the application running as illustrated by the application dashboard described in Sect. 5.4.

5.2 MALP Architecture

In order to obtain fine-grained performance data, tracing approaches move analysis to offline. Our work investigates whether it is possible to reduce the impact on the application of online analysis through network-coupling of computing resources dedicated to online data projection. In this section, we detail the steps which turned MALP, designed for online reduction [4, 5], into an in-situ analysis engine.

As described in [5], on the analysis side, we implemented a data-flow engine inspired from the *Blackboard* expert systems, the idea being to have several (knowledge system) plug-ins valorizing incoming performance events in an orthogonal manner by registering themselves on a given event type. In MALP most of these analysis are performing either temporal or functional reductions. The interprocess aspect was originally handled by synchronous MPI reductions when application ended, reducing data accumulated in each module. Due to this intermediate MPI step, we were not taking advantage of the online nature of MALP as unified performance data were only reachable at the end of the execution in a PDF report. The spatial reduction acting as a barrier on our data-collection process, we looked for an alternative design in order to allow runtime interactions. Following the Blackboard idea, we then changed MALP design to integrate a visualization agent interacting with distributed processes. The idea being that reduction plug-ins would produce data which could then be directly consumed by rendering components.

As shown in Fig. 5.2, we added a new visualization agent. It was implemented in JavaScript on top of *Node.js* in order to be modular while taking advantage of built-in networking libraries (TCP and HTTP). In this new infrastructure, each plug-in describes a template-based visualization component which is loaded inside

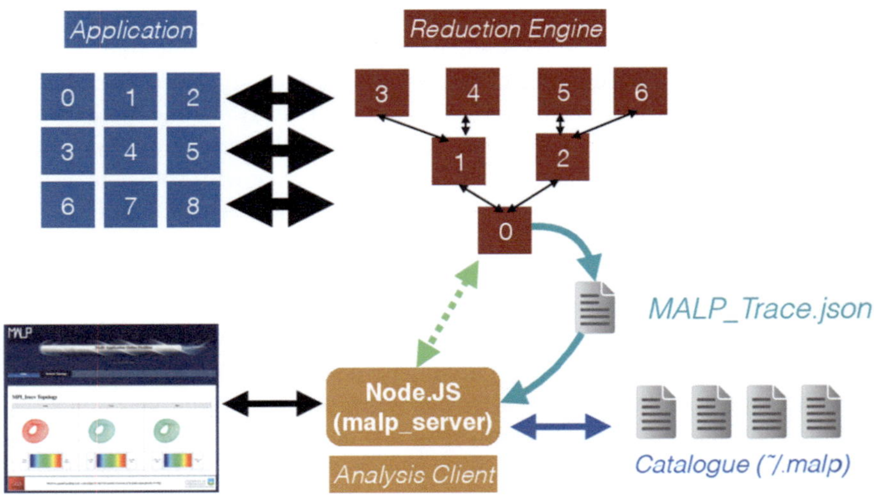

Fig. 5.2 MALP instrumentation chain including the web-based rendering component

a lightweight web-server. Server distributing dynamic web-pages to client web-browsers, opening the way for interactive data-exploration (further described in Sect. 5.3). The coupling between the application and the analysis engine remained unchanged, forwarding fine-grained events. Dealing with the coupling between the server and the analysis engine we decided to adopt a shared data-model in order to provide generic storage capabilities. Our server being in JavaScript and our target rendering environment being a web-browser (also JavaScript enabled), we decided that the plug-ins located in the reduction engine would output *JavaScript Object Notation (JSON)* [7] formated data. As the analysis was implemented in C, we defined serializable C data-structures allowing analyses to store arbitrary data in a hierarchy of JSON objects. Moreover, data for each module is stored inside a common root object, allowing the whole performance data-set to be serialized at once. This root JSON object, storing arbitrary sub-objects defined a common data-structure for performance data, this with almost no constraints unlike in trace-formats—what is stored is at the discretion of the plug-in.

As shown in Fig. 5.2, to deal with the coupling between the analysis engine and the Node.JS web-server, we relied on two methods. The first one uses a JSON file which is obtained after performing a reduction of the per-process performance data in the root analysis task—approach similar to what was previously done for our PDF reports. The JSON file is generated by serializing the root JSON object which contains the data from all plug-ins. Such files can be stored in a performance catalog for later retrieval in the server. The second coupling approach is an online one. Thanks to a TCP socket connected to the root of a binary tree gathering analysis tasks, the web-server is able to send and query JSON objects, relying on serialization and deserialization primitives in both C and JavaScript. We call this coupling the *Node Shared Cache*, making it possible for the C data located in a remote process to be queried from the Node.JS web-server and transitively from a web-browser.

To illustrate MALP's ability to couple performance data with analysis, the following sections describes our new performance-data visualization (Sect. 5.3) and interactive application control (Sect. 5.4).

5.3 Profiling with MALP

In this section we briefly illustrate MALP profiling capabilities. The new web-based interface, replacing a PDF report, allowed the implementation of original performance visualizations, taking advantage of web technologies (D3.js, WebGL, plot libraries …) to provide an interactive user-friendly interface while minimizing development cost. MALP now provides a wide range of performance analysis modules with, for example, profiles, temporal and spatial behavior maps, load-balancing charts in Fig. 5.3a and interactive communication topologies in Fig. 5.3b. JSON profile files generated by the analysis engine can be directly loaded through the web-interface, being stored in a *performance catalog* for latter reference.

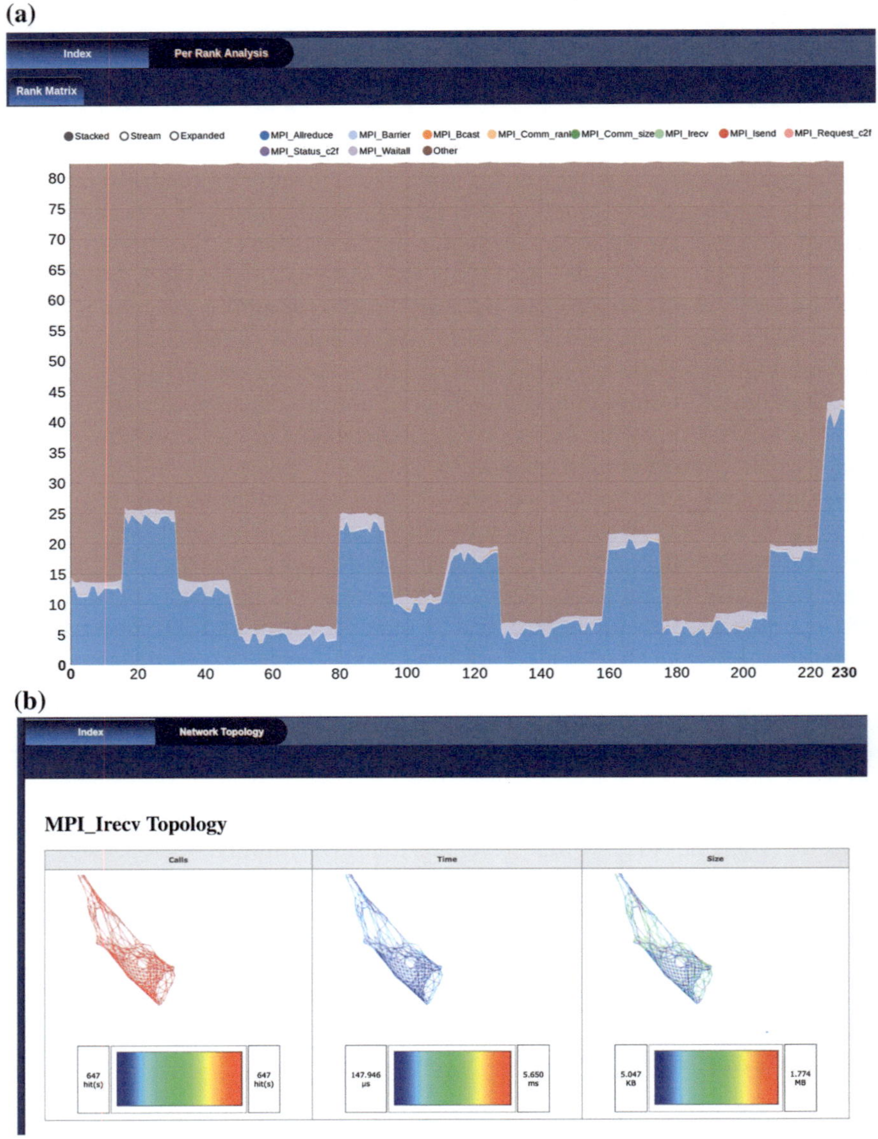

Fig. 5.3 Screenshots from MALP web-interface. **a** Process view of an imbalanced computation. **b** Interactive 3D topology viewer

The templating system developed in the Node.JS part of MALP allows the expression of performance analysis in a straightforward manner, the JSON object exported by the analysis engine being directly addressable programmatically by the JavaScript abstraction. This process can be illustrated with Fig. 5.3a presenting MPI profile balancing and generated from the following per-task MPI profile:

```
[[{"hits":128,"size":1468010000,"name":"MPI_File_read","time":3.65},..],..]
```


This JSON array is first indexed with the MPI rank and then presents a per-rank profile gathering for each instrumented function, the number of calls, the cumulative time and the total size when suitable. From this point, generating the MPI balancing graph was then just a matter of converting the JSON data-layout to match the requirements of the plotting library (here NVD3). Similarly, if we consider Fig. 5.3b, we also start from exported JSON data providing connectivity information for each MPI point-to-point in terms of hits, time and size as follows:


```
{"MPI_Recv":[{"src":1,"hits":4,"size":192,"dest":0,"time":0.0005},..],..}
```


The topology analysis, uses these data to generate a Graphviz [11] graph description file in order to compute the three-dimensional layout of the graph. Then Graphviz's output, enriched with coordinates is reloaded in the server, coordinates are centered around object's center of gravity and normalized before being passed to the interactive three.js WebGL layer inside the client web-browser.

These two simple examples illustrate the advantage of our JSON approach which is directly operated from the JavaScript layer for both pre-processing and rendering in a web-browser. If compared to a binary trace format, this approach is clearly less space efficient but makes data-analysis much simpler. Moreover, adding new events or plugins to the reduction engine requires no modification in the intermediate format unlike trace formats with (in general) a constrained event set. However, the interface was not yet taking advantage of the collocation with the running application, inciting us to explore new use cases exemplified in next sections.

5.4 Introducing Parallel Application Dashboards with MALP

The *Node Shared Cache* is also a way of interacting with running applications, in this section we illustrate its use to monitor a sample parallel code. The motivation for this is that data-management techniques developed for performance monitoring can clearly be applied more globally to application data-management.

Figure 5.4 presents an HTML interface that we designed in a few hours on top of MALP online coupling facilities. We made a simple program computing Pi by integrating *Arctan* with the Euler approximation method (as known as the rectangle method). In Fig. 5.4a, we use the HTML page hosted in the Node.JS server to first explain our numerical method relying on $\int_0^1 Arctan(x)\mathrm{d}x = \frac{\pi}{4}$. Then we allow the user to *interactively* set the number of intervals (or rectangles) to be used, before clicking on compute. This has the effect of sending a GET operation to the server which then relays the information to the MPI processes by setting a value in the *Node Shared Cache*—parallel computation being triggered on value change. Then, when the processes are done computing the integral on their own sub-intervals,

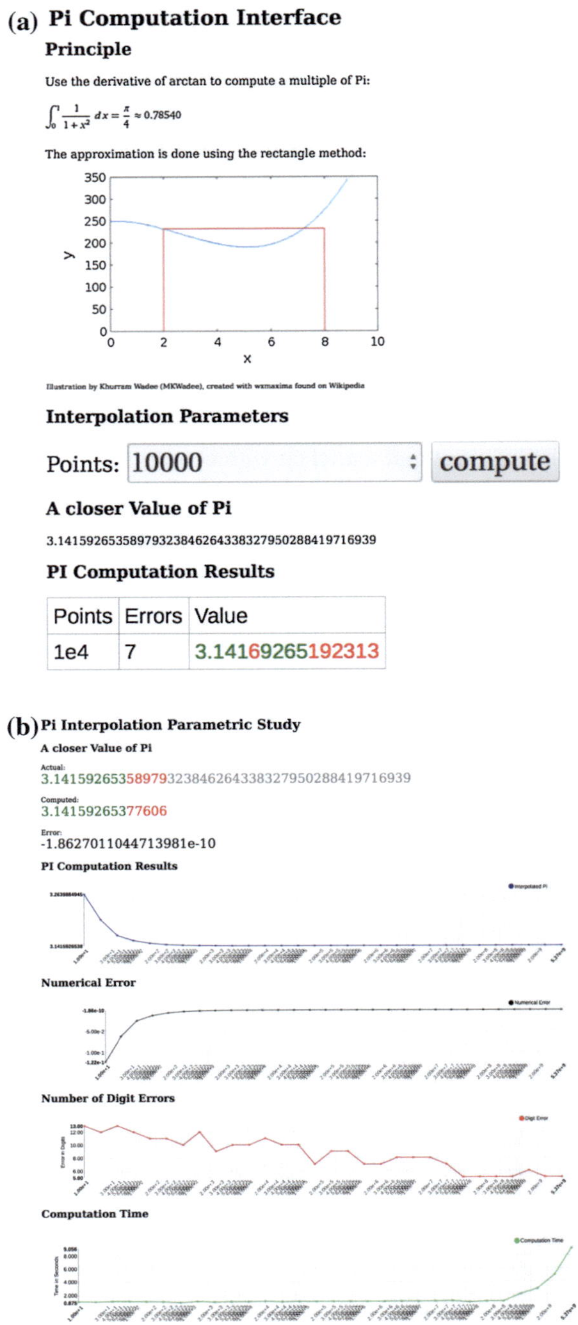

Fig. 5.4 Overview of our on-line application management interface for π computation. **a** Interactive computation interface. **b** Computation dashboard

rank zero stores the approximated π value in the cache, event which is notified in the server. Eventually, on client side, a JavaScript function periodically queries current value, displaying it in a color-coded fashion with correct digits in green and bad ones in red. This first example demonstrated bidirectional online coupling with a running application. The second example of Fig. 5.4b uses a similar approach, however, instead of waiting for the user to input a number of rectangles, a script in the browser does a parametric analysis with an increasing number to study the convergence towards π. In complement of the colored value, we present dynamic curves respectively displaying successive approximations, relative error, number of false digits and computation time.

This example, clearly out of the scope of a profiling tool was chosen to illustrate the potential shifting that exists from complex data-management approaches developed for profiling to the more general problematic of managing code outputs. We shown that starting from our online profiling engine we were able to implement with a reduced development effort an interface providing new interaction opportunities with running applications. We are convinced that hybrid data-management techniques will play an important role in the usability of Exascale machines. Justifying in our opinion the addition of primitives allowing such behavior in runtimes or programming interfaces such as MPI as we further detail in the conclusion.

5.5 Conclusion an Future Work

As discussed in Sect. 5.1.5, we think that interactions with parallel applications will benefit from a *multi-scale* approach, combining all the coupling paradigms that we covered in our introductory content. To do so, suitable mechanisms have to be defined at programming interface level in order to pave the way for such online interactions. We are currently working on MALP capabilities transposition as extensions to the MPI standard.

Dealing with MPI virtualization [5, 22], the standard is already able to join disjoint MPI applications using MPI_Comm_connect, creating an inter-communicator with the exact same features than `MPI_COMM_UNIVERSE`. It is then possible to achieve the virtualized MPI behavior with current standard, connecting several applications or services. As far as the data-stream interface is concerned, it can also be implemented inside MPI as we already did it using only standard calls [5]. However, one point that we see as missing in MPI is *query* support. Indeed, for performance reason, MPI mostly provides either paired communications or RDMAs but no easy manner to implement queries which could be modeled as Remote Procedure Calls (RPC). For us, the ability to *opportunistically* interact with a remote process is a key component. Using current standard, one would need to have a loop doing MPI_Recv, `ANY_SOURCE` to process incoming request which would then be processed by an arbitrary function before returning a result. To do so, the application has to handle a pool of threads processing such messages while making sure that the underlying runtime supports the `THREAD_MULTIPLE` level. If handled by the runtime the use of

RPCs would become more convenient and more importantly standardized between applications, allowing transversal tool services definition.

Such interface has already been investigated in the MPI context and called *active messages* [8], such messages were implemented in AMMPI and AM++ [25] but not standardized yet. One could imagine with such interface, querying performance data from a given process, sending a command to a simulation, initiating a two sided communication phase by creating the remote context through an RPC for in-situ visualization, load-balancing tasks on a set of nodes and gathering results, ... In this purpose, we propose a similar RPC interface for *service* abstraction, going a bit further than Active Messages [18] by proposing a global naming (URI and domain):

```
/* Emit a Query to a remote process or a global URI */
MPIX_Query( int domain, char * uri,
            void * post_buff, size_t post_size,
            void ** get_buff, size_t ** get_size,
            MPI_Communicator comm, MPI_Request * req );
/* Register a function to fulfil global or local queries at a
   given URI */
MPIX_Bind( int domain, char * uri,
           int (*mpi_rpc)( void * post_buf,
                           size_t post_size,
                           MPI_Query * handle ));
/* Return a result using a query context */
MPIX_Query_ret(  MPI_Query * handle,
                 void * get_buff,
                 size_t get_size );
/* Return an error code using a query context */
MPIX_Query_error(  MPI_Query * handle,  int code );
```

This interface would provide query support in MPI through the MPIX_Query function call. Either locally or globally using respectively the `MPI_SCOPE_LOCAL` and `MPI_SCOPE_GLOBAL` keywords as `domain` parameter. The `uri` defines the resource to be queried (just like in an HTTP request) and the other parameters define both what is sent in order to be passed as argument in the remote function and the buffer which is returned (note that it is allocated) from the call. Eventually the query can be waited as usual with a request. In order to expose an RPC service, a process has to call the MPIX_Bind function which takes a `domain` argument (local or global), an `uri` to be matched by the calling RPC and a function associated with the RPC. Eventually, we introduce two functions to be used to return a result from an RPC, to do so, they rely on the `MPI_Query` context passed in argument of the RPC handler. The function can then either return an arbitrary buffer using MPIX_Query_ret or an error code (reported on the caller) using MPIX_Query_error.

This small MPI interface extension, is what we consider to be an important component to enable more orthogonality between applications, exposing several data-sources even from pre-loaded libraries through the same query interface. This leads to a more orthogonal vision of an MPI computation with a convenient expression of the master–slave/client–server model which is not present yet in the standard being mostly used for SPMD computation. Thanks to this data-model, it will be much easier to have several MPI applications collaborating around *services* which could be exposed using this interface through an horizontal vision of the parallel computation with IOs, visualization, profiling, logging, output-management and computation processes collaborating in an online fashion, as a *constellation of services*.

References

1. Adhianto L, Banerjee S, Fagan M, Krentel M, Marin G, Mellor-Crummey J, Tallent NR (2010) HPCToolkit: tools for performance analysis of optimized parallel programs. Concurr. Comput.: Pract. Exp. 22(6):685–701
2. Arnold, D.C., Ahn, D.H., de Supinski, B.R., Lee, G.L., Miller, B.P., Schulz, M.: Stack trace analysis for large scale debugging. In: IEEE International Parallel and Distributed Processing Symposium, IPDPS. pp. 1–10. IEEE (2007)
3. Benedict, S., Petkov, V., Gerndt, M.: Periscope: an online-based distributed performance analysis tool. Tools for High Performance Computing 2009, pp. 1–16. Springer, Berlin (2010)
4. Besnard, J.B.: Profiling and Debugging by Efficient Tracing of Hybrid Multi-Threaded HPC Applications. Ph.D. thesis, Université de Versailles Saint Quentin en Yvelines (2014)
5. Besnard, J.B., Pérache, M., Jalby, W.: Event streaming for online performance measurements reduction. In: 42nd International Conference on Parallel Processing (ICPP), pp. 985–994. IEEE (2013)
6. Chan A, Gropp W, Lusk E (2008) An efficient format for nearly constant-time access to arbitrary time intervals in large trace files. Sci. Program. 16(2–3):155–165
7. Crockford, D.: The Application/Json Media Type for Javascript Object Notation (json) (2006)
8. von Eicken, T., Culler, D.E., Goldstein, S.C., Schauser, K.E.: Active messages: a mechanism for integrated communication and computation. In: Proceedings of the 19th Annual International Symposium on Computer Architecture, ISCA '92, pp. 256–266. ACM, New York, NY, USA (1992). http://doi.acm.org/10.1145/139669.140382
9. Eschweiler D, Wagner M, Geimer M, Knüpfer A, Nagel WE, Wolf F (2011) Open trace format 2: the next generation of scalable trace formats and support libraries. PARCO. 22:481–490
10. Frings, W., Wolf, F., Petkov, V.: Scalable massively parallel i/o to task-local files. In: Proceedings of the Conference on High Performance Computing Networking, Storage and Analysis, pp. 1–11. IEEE (2009)
11. Gansner ER, North SC (2000) An open graph visualization system and its applications to software engineering. Soft. -Pract. Exp. 30(11):1203–1233
12. Geimer M, Wolf F, Wylie BJ, Ábrahám E, Becker D, Mohr B (2010) The scalasca performance toolset architecture. Concurr. Comput. Pract. Exp. 22(6):702–719
13. Hilbrich, T., Müller, M.S., de Supinski, B.R., Schulz, M., Nagel, W.E.: GTI: A generic tools infrastructure for event-based tools in parallel systems. In: IEEE 26th International Parallel & Distributed Processing Symposium (IPDPS), pp. 1364–1375. IEEE (2012)
14. Hilbrich, T., Schulz, M., de Supinski, B.R., Müller, M.S.: MUST: A Scalable approach to runtime error detection in MPI programs. Tools for High Performance Computing 2009, pp. 53–66. Springer, Berlin (2010)
15. Knüpfer, A., Brendel, R., Brunst, H., Mix, H., Nagel, W.E.: Introducing the open trace format (OTF). Computational Science–ICCS 2006, pp. 526–533. Springer, Berlin (2006)
16. Knüpfer, A., Brunst, H., Doleschal, J., Jurenz, M., Lieber, M., Mickler, H., Müller, M.S., Nagel, W.E.: The vampir performance analysis tool-set. Tools for High Performance Computing, pp. 139–155. Springer, Berlin (2008)
17. Knüpfer, A., Rössel, C., an Mey, D., Biersdorff, S., Diethelm, K., Eschweiler, D., Geimer, M., Gerndt, M., Lorenz, D., Malony, A., et al.: Score-P: a joint performance measurement run-time infrastructure for periscope, scalasca, TAU, and vampir. Tools for High Performance Computing 2011, pp. 79–91. Springer, Berlin (2012)
18. Mainwaring, A.M., Culler, D.E.: Active message applications programming interface and communication subsystem organization. Technical Report UCB/CSD-96-918, EECS Department, University of California, Berkeley (Oct 1996). http://www.eecs.berkeley.edu/Pubs/TechRpts/1996/5768.html
19. Nataraj, A., Malony, A.D., Morris, A., Arnold, D., Miller, B.: A framework for scalable, parallel performance monitoring using TAU and MRnet. In: International Workshop on Scalable Tools for High-End Computing (STHEC 2008), Island of Kos, Greece (2008)

20. de Oliveira Stein, B., de Kergommeaux, J.C., Mounié, G.: Pajé Trace File Format. Technical report, ID-IMAG, Grenoble, France, 2002. http://www-id.imag.fr/Logiciels/paje/publications (2010)
21. Roth, P.C., Arnold, D.C., Miller, B.P.: MRNet: A Software-based multicast/reduction network for scalable tools. In: Proceedings of the 2003 ACM/IEEE Conference on Supercomputing, p. 21. ACM (2003)
22. Schulz, M., de Supinski, B.R.: P^nMPI tools: a whole lot greater than the sum of their parts. In: Proceedings of the 2007 ACM/IEEE Conference on Supercomputing, p. 30. ACM (2007)
23. Shende SS, Malony AD (2006) The TAU parallel performance system. Int. J. High Perform. Comput. Appl. 20(2):287–311
24. Vetter, J., Chambreau, C.: MPIP: Lightweight, scalable MPI profiling (2005)
25. Willcock, J.J., Hoefler, T., Edmonds, N.G., Lumsdaine, A.: AM++: a generalized active message framework. In: Proceedings of the 19th International Conference on Parallel Architectures and Compilation Techniques, PACT '10, pp. 401–410. ACM, New York, NY, USA (2010). http://doi.acm.org/10.1145/1854273.1854323
26. Zaki O, Lusk E, Gropp W, Swider D (1999) Toward scalable performance visualization with jumpshot. Int. J. High Perform. Comput. Appl. 13(3):277–288

Chapter 6
Evaluation of Tool Interface Standards for Performance Analysis of OpenACC and OpenMP Programs

Robert Dietrich, Ronny Tschüter, Tim Cramer, Guido Juckeland and Andreas Knüpfer

Abstract A fundamental interest in application development for high performance computing (HPC) is a close-to-optimal execution efficiency. To systematically achieve this, it is reasonable to use performance analysis tools that provide an insight into the execution of a program. Programming models that also specify tool interfaces enable the design of robust and portable tools. This paper investigates current efforts in the specification of performance tool interfaces for the parallel programming standards OpenACC and OpenMP. Technical reports have been published about an OpenACC and an OpenMP tools interface. We integrated both interfaces in the measurement infrastructure Score-P and explored their applicability for instrumentation-based performance analysis. We also consider recent extension proposals for the portable collection of performance data on the device for OpenMP and OpenACC.

6.1 Introduction

Node-level parallelism in today's supercomputers emerges from multi-core processors and many-core devices. The latter ones are accelerators based on graphics processing units (GPGPUs) or Intel Xeon Phi coprocessors. This heterogeneous composition of computing resources creates the need to utilize multiple parallel programming paradigms in a single application, e.g. multi-threading on the CPU and computation offloading to a many-core device.

R. Dietrich (✉) · R. Tschüter · G. Juckeland · A. Knüpfer
Center for Information Services and High Performance Computing, Technische Universität Dresden, 01062 Dresden, Germany
e-mail: robert.dietrich@tu-dresden.de

T. Cramer
IT Center, RWTH Aachen University, 52074 Aachen, Germany

T. Cramer
Chair for High Performance Computing, RWTH Aachen University, 52074 Aachen, Germany

T. Cramer
JARA - High-Performance Computing, Schinkelstraße 2, 52062 Aachen, Germany

A. Knüpfer et al. (eds.), *Tools for High Performance Computing 2015*,
DOI 10.1007/978-3-319-39589-0_6

Directive-based programming standards such as OpenACC and OpenMP are widely used in high performance computing to simplify the programming and effectively use these resources without the need for complex low-level APIs. The OpenACC standard has established as an effective alternative to offloading models such as CUDA and OpenCL. OpenMP has initially been designed for parallel programming on shared memory systems. With the introduction of device directives in version 4.0 of the specification, it can also be used to offload computation to accelerators or coprocessors that may have a separate memory address space [3].

Both programming models, OpenACC and OpenMP, hide complex details that are specific to the underlying hardware architectures. However, the application developer can still explicitly specify details such as data management and work-sharing to utilize the available resources more efficiently. To understand the performance of an application that uses directive-based parallelization it is necessary to obtain information about the execution of directives including the implicit details that have not been specified by the programmer. OpenACC and OpenMP specify compiler directives, which in contrast to host functions are commonly not visible with compiler-based instrumentation techniques. Therefore, it emerges the need for standardized interfaces that enable performance tools to investigate the implementation and the execution of such directives.

This paper discusses performance analysis for OpenACC and OpenMP, based on their most recent tool interface proposals. The OpenACC consortium recently released a version of the OpenACC standard 2.5 [15] for public feedback which includes a revised version of the tool interface that has been previously published as technical report [13]. The OpenMP architecture review board (ARB) published a technical report about the OpenMP tools (OMPT) interface [5] that has been designed by the OpenMP tools working group and is targeted to be a part of the OpenMP 5.0 standard. A working document on OMPT with recent changes is publicly available [7] and included in our discussion. We evaluate the benefit and the applicability of both interfaces for performance analysis. In addition, the functionality and concept of OMPT and ACCT will be compared based on an implementation in the Score-P analysis infrastructure.

The remainder of this paper is organized as follows: Sect. 6.2 describes background information about the programming standards OpenACC and OpenMP. In Sect. 6.3 we present the OpenACC tools (ACCT) interface [13] and in the following section (Sect. 6.4) we depict the OpenMP tools (OMPT) interface [6]. An evaluation and comparison of both interfaces is performed in Sect. 6.5. Finally, in Sect. 6.6, we present our conclusions.

6.2 OpenACC and OpenMP

Directive-based programming standards such as OpenACC and OpenMP provide an abstraction that hides hardware specifics. Additionally the abstraction targets interoperability across different hardware and compiler vendors.

Both programming standards support the programming languages C, C++ and Fortran. OpenACC as well as OpenMP include offloading in their execution model. They let the programmer annotate compute regions so that their execution will be offloaded to an accelerator device. The host-side—the CPU—retains the control. We call this model *host-directed*. Data lifetime, data transfer between host and device as well as host-device synchronization are also host-directed.

6.2.1 OpenACC

The OpenACC standard [15] has been initiated to simplify the programming of high-level host and accelerator codes without the need to explicitly initialize the accelerator, manage data or control transfers between host and accelerator. The OpenACC memory model distinguishes devices that physically share memory with the host and devices that have a separate memory. For the latter case, the device data environment will be shared with the local thread and an OpenACC runtime implementation does neither need to create copies of data nor transfer data to the device.

The *data* construct, *enter data*, and *exit data* directives are used to control data lifetime on the accelerator as well as data movement between host and accelerator. Data regions can also be implicitly generated for a compute construct when it is annotated with an explicit data clause. The *update* directive is used to update data on the device with its corresponding data on the host or vice versa.

OpenACC specifies two fundamental compute constructs, *kernels* and *parallel*, to offload the execution of an annotated code block to an accelerator. The *kernels* construct defines a region in the program that will be executed on the accelerator device as a sequence of kernels. According to the compiler's capabilities it will generate code that will be executed more or less efficiently on the device. OpenACC distinguishes gang, worker and vector parallelism. A gang is a group of workers and maps to a CUDA thread-block or OpenCL work-group. Workers may have vector execution capabilities and map to CUDA threads or OpenCL work-items. "Vector parallelism is for SIMD or vector operations within a worker [15]". Typically, loop nests will be compiled into separate kernels which do not necessarily need to run with the same gang, worker and vector size.

The *parallel* construct defines a region in the program that will be executed in parallel on the device by each gang. Outside of a gang-parallel *loop* construct, which must be immediately followed by a loop, all gangs execute the code redundantly. Gang, worker and vector size remain constant during the execution of the *parallel* region.

As with OpenMP, OpenACC clauses allow the programmer to explicitly specify the type of parallelism for individual compute constructs. Data clauses explicitly manage device data or control the data movement direction.

6.2.2 *OpenMP*

OpenMP is a widely used parallel programming standard for shared-memory platforms. Many compilers support it to exploit thread-level parallelism on modern hardware architectures. It allows users to express parallelism in a program by introducing directives that extend the base language with different constructs, e.g. tasking, work-sharing and synchronization constructs. Due to the support for offloading code sections to accelerator devices in OpenMP 4.0 the traditional model for shared memory was extended. In current heterogeneous systems the accelerator devices do not have the same address space as the host system. Therefore, it is necessary to transfers data to or from a device when code sections are offloaded to accelerator devices. OpenMP 4.0 provides functionality for data transfers and ensuring data consistency between host and target device. However, in future systems devices might share the same address space as the host. Hence, OpenMP calls the corresponding clause *map* instead of *copy* as in OpenACC. The device model mainly consists of the constructs *target* to offload a code region, *target data* to create a device data region and *target update* to make data in the device environment consistent with the host.

6.3 The OpenACC Tools Interface

The OpenACC consortium has recently published a technical report [13] that specifies a performance tools interface for profiling and tracing. It enables the recording of OpenACC runtime events that occur during the execution of an OpenACC program and is therewith intended as an interface for instrumentation-based performance tools. A tool only needs to register for event types and implement respective callback routines. An OpenACC runtime will call these specified routines when a respective event has been triggered. The ACCT interface also defines an integration for low-level programming models such as CUDA and OpenCL to gather additional information that is not provided by an OpenACC runtime. OpenACC runtime states for sampling-based tools and a portable interface for device data collection are proposed in [4]. PGI compilers of version 15 and above provide a test implementation of the ACCT interface that includes most features.

6.3.1 *Runtime Events*

The ACCT interface defines three groups of events: data events, launch events and other events. Data and launch events provide additional details about the respective operation. For data events, an OpenACC runtime might expose the name of the variable, the number of bytes, a host pointer, and a device pointer to the corresponding data. Launch events are associated with a kernel name as well as the gang, worker

Table 6.1 OpenACC runtime events specified in [15] (updated and revised from [4])

Event(s) acc_ev_	Occurrence
Launch events	
enqueue_launch_[start|end]	Before/after a kernel launch operation
Data events	
enqueue_upload_[start|end]	Before/after a data transfer to the device
enqueue_download_[start|end]	Before/after a data transfer from the device
create	When the OpenACC runtime associates device memory with host memory
delete	When the OpenACC runtime disassociates device memory from host memory
alloc	When the OpenACC runtime allocates memory from the device memory pool
free	When the OpenACC runtime frees memory from the device memory pool
Other events	
device_init_[start|end]	Before/after device initialization
device_shutdown_[start|end]	Before/after device finalization
compute_construct_[start|end]	Before/after the execution of a compute construct
update_[start|end]	Before/after the execution of an update directive
enter_data_[start|end]	Before/after the execution of an enter data directive, the entry to data constructs and implicit data regions
exit_data_[start|end]	Before/after the execution of an exit data directive, the exit from data constructs and implicit data regions
wait_[start|end]	Before/after an OpenACC wait operation
runtime_shutdown	Before the OpenACC runtime finalizes

and vector dimension size. Available events are listed in Table 6.1. Most events are reported in pairs of a start and an end event. Data allocation and deallocation events are atomic and it is not specified when exactly they are triggered (before, after, or during the activity). The event for runtime shutdown is also atomic and triggered before the OpenACC runtime shuts down. This allows tools to finish actions that rely on the OpenACC runtime. Events on implicit and explicit wait operations can be used to identify wait states in the execution of an OpenACC program. Low-level accelerator runtime events can be gathered via the respective low-level interface, which is independent of the ACCT interface. However, it is possible to correlate information between interfaces such as for example accelerator kernels by their names.

6.3.2 Callback Signature and Low-Level API Interface

The signature of event callbacks has three arguments and is the same for all event types. The first argument is a pointer to a structure that contains general information such as the interface version, the OpenACC device type, the host thread ID and the source code location of the construct or directive that triggered the event.

The second argument is the event-specific information, defined as a union with the first field being the event type. Depending on the event type, fields of the respective union member contain valid event-specific information. The first three fields of each union member are identical: event type, number of valid bytes in the respective structure, and the *toolinfo* field. The *toolinfo* field can be used to pass information from a start event to the corresponding end event. As described in Sect. 6.3.1 there are three groups of events. According to the event groups the ACCT interface defines three union members that provide information about data events, launch events or other events.

The third argument in each callback provides information about the low-level API of the target device to gather details that are not provided with the ACCT interface itself. Furthermore, it enables tools to hook into the low-level programming model execution and for example insert profiling events or collect other performance-relevant information. Considered target APIs are CUDA and OpenCL, but they might also be Intel Coprocessor Offload Infrastructure (COI) or others. One field in this structure specifies the vendor and three fields are pointers to API-specific handles for the device, the context and the asynchronous queue. CUDA and OpenCL have logical counterparts for these handles which are the types `CUdevice`, `CUcontext` and `CUstream` for the CUDA driver API and `cl_device`, `cl_context` and `cl_command_queue` for OpenCL.

6.3.3 Tool Registration, Initialization and Control

A tool library that implements the ACCT interface has to define an initialization routine `acc_register_library`. It is linked either statically or dynamically with the application. The OpenACC runtime or the application itself calls the initialization routine. An OpenACC runtime may also support to load a tool's library dynamically by reading the environment variable `ACC_PROFLIB` or `LD_PRELOAD` (on Linux systems) specifying the path to the tool's library.

The ACCT API also provides a mechanism to register and unregister events as well as temporarily disable and enable callbacks. Pointers to the registration routines `acc_prof_register` and `acc_prof_unregister` are passed to the initialization routine `acc_register_library`. The first two arguments are the event type for which a callback is being registered and a pointer to the respective callback routine. The third argument is zero for registering and unregistering event callbacks.

To temporarily disable or re-enable a callback routine for an event, `acc_toggle` can be passed as third argument. There is no restriction to when events can be registered, unregistered, enabled, or disabled.

6.4 OMPT—An OpenMP Tools Interface

There were previous efforts on the specification of a tool interface. For example, the Sun/Oracle collector API [8] was designed for sampling-based tools. In contrast, the POMP2 interface [11], with a respective implementation in the OPARI2 source-to-source instrumentor, supports instrumentation-based tools. Other frameworks as for example the ROSE compiler [9] also allow to instrument OpenMP applications. Unfortunately they failed to become part of the OpenMP specification, but inspired the design of the OMPT interface.

The OMPT interface as specified in the published technical report [5] has been designed as a portable interface for both sampling-based and instrumentation-based performance measurement tools. The goal of this technical report is to be part of the OpenMP 5.0 specification, which is expected to be released in 2017 or 2018. Since the OpenMP specification itself develops over time, the technical report is also revised continuously [7]. In this section we cover the specification of both versions of the document and refer to them as *official* technical report, which was published by the OpenMP Architecture Review Board (ARB), and as *revised* technical report, which is under development of the OpenMP tools subcommittee.

OMPT defines an interface to obtain information from the OpenMP runtime system during the execution of an OpenMP program. Initial implementations of OMPT are available as part of IBMs lightweight OpenMP runtime system (less than 1 % runtime overhead for OMPT) and Intels open-source OpenMP runtime library as mentioned in [6]. The official technical report on OMPT is an extension proposal for OpenMP 3.1 and was planned to be part of an upcoming OpenMP specification. It does not consider features that have been introduced in version 4.0 of the OpenMP specification (e.g. device directives and task dependencies).

6.4.1 Runtime States

The OMPT interface provides access to state information of the OpenMP runtime that can be queried by an analysis tool. This information is maintained by the OpenMP runtime for each thread and approximates the current state of the thread. Sampling-based performance analysis tools can use the different states and the corresponding state transitions of an OpenMP application during the execution to analyze an application's runtime behavior. There are three classes of states in the official technical report: mandatory, optional, and flexible.

A compliant OpenMP runtime with OMPT support must implement the five mandatory states and has to differentiate between a thread waiting to execute an OpenMP parallel region (`ompt_state_idle`), a thread executing code outside all parallel regions (`ompt_state_work_serial`), and a thread executing code in a parallel region (`ompt_state_work_parallel`). Another mandatory state exists for a thread that is neither a user thread, nor an initial thread, nor a thread that is not (yet) part of an OpenMP team (`ompt_state_undefined`). The last mandatory state is a placeholder, which can be used to enumerate all available states and will never be reported by a runtime.

The two optional states may or may not be implemented in a standard-compliant OpenMP runtime. They report that a thread is combining partial reduction results (`ompt_state_work_reduction`) or expose runtime overheads like preparing a parallel region or a new explicit OpenMP task (`ompt_state_overhead`). The nine flexible states report that a thread is waiting at any kind of barrier, lock, taskwait, taskgroup, critical, atomic or ordered construct. They are called flexible since an OpenMP runtime may decide when it exactly switches the state of a thread. This might happen early when the thread encounters the construct, or late when a thread begins to wait to enter the corresponding region.

In the revised OMPT document the states are not explicitly divided into these three classes any more. The defined states are the same, but a standard-compliant runtime has to implement all of them.

6.4.2 Runtime Events and Callbacks

The OMPT interface enables instrumentation-based performance tools to register function callbacks for events of interest. Event callbacks are classified as mandatory or optional. The set of eight mandatory events has to be implemented by a standard-compliant OpenMP runtime. It contains begin/end event pairs for threads, parallel regions and explicit tasks. Furthermore, one event for the application tool control and one event for the runtime shutdown are mandatory. The revised technical report [7] extends this list with an additional begin/end event pair for target tasks. A performance tool can only rely on this minimal set of events as common functionality in all OpenMP runtimes.

The optional events enable tools to gather and analyze more detailed information about an OpenMP program. They are divided into two different sets: Events for blame shifting and events for instrumentation-based measurement tools. The former set contains event pairs which allow to measure the time spent idling outside a parallel region, in a barrier, in a taskwait or taskgroup region or in any kind of lock (e.g., OpenMP API lock, critical region etc.). Thus, it can be used to shift the costs of idling within an application or the OpenMP runtime from symptoms to causes. The latter event set enables instrumentation-based tools among others to gather and analyze implicit and initial task creation and destruction, lock creation and destruction or begin/end event pairs for loop, section or barrier constructs. Even if an OpenMP runtime does not implement any of these optional events, it remains standard-compliant. In this way, the OMPT design objective to not impose unreasonable implementation burdens on the runtime developer is fulfilled.

6.4.3 Tool Registration, Initialization and Control

In order to use the OMPT interface a tool must register itself during the initialization by providing an implementation of the `ompt_initialize` function. This function will be called by the OpenMP runtime immediately after the runtime is initialized. The first parameter passed to the initialization function is a lookup callback (`ompt_function_lookup_t`), which must be used to obtain function pointers to the OMPT inquiry functions. Inquiry functions are among others used to retrieve data from the execution environment (e.g., `ompt_get_thread_id`) or to register an event (`ompt_set_callback`). A tool uses the latter one to get triggered for a respective event. The lookup mechanism hides OMPT-specific functions that are not visible as global symbols in an OpenMP runtime library and therefore cannot be called directly. In order to control the tool initialization the environment variable `OMP_TOOL` is used. If it is set to *enabled*, but no tool has attached by providing a version of `ompt_initialize`, a weak symbol version of `ompt_initialize` will be provided by the runtime. If `OMP_TOOL` is set to *disabled* `ompt_initialize` will not be invoked at all.

An application can control an attached tool to start, stop, pause, or restart monitoring with the routine `ompt_control`. A tool can also define additional command codes. Currently, the OMPT interface does not provide a way for tools to enable, disable, register or unregister event callbacks at runtime after the initialization.

6.5 Evaluation

OpenMP has initially been designed for multi-threading on shared-memory systems, but it also incorporates tasking and offloading in its latest specification [14]. OpenACC has been designed to facilitate computation offloading to accelerators and therewith covers only one paradigm that OpenMP supports. Despite both standards cover a different set of paradigms and features their concepts for integrating tool interfaces can be compared. The ACCT interface will be part of the OpenACC 2.5 standard [15], whereas the OMPT interface is only planned to be part of OpenMP 5.0, while OpenMP 4.5 is not yet released.

In the following, we compare the design of the OMPT and the ACCT interface and evaluate the performance data collection approaches according to functionality and features as well as the implementation burden for tool developers. We study the integration into the performance measurement infrastructure Score-P and describe the interaction of tool and runtime library via the interfaces. Finally, we highlight differences and similarities of both interfaces.

6.5.1 Interface Design

The presented tool interfaces OMPT and ACCT define a portable way to collect performance data from OpenMP and OpenACC runtimes. Figure 6.1 illustrates the interaction between a tool and an OpenMP/OpenACC runtime library. The initialization mechanism is similar in both interfaces. The performance tool implements a defined initialization routine (OMPT: `ompt_initialize`, ACCT: `acc_register_library`), which will be called by the respective runtime. The initialization routine provides arguments which enable a tool to obtain pointers to event registration routines and for OMPT also to inquiry functions. OMPT uses an additional indirection via the lookup routine (see Sect. 6.4.3). ACCT-enabled runtimes directly pass pointers to the event registration routines (see Sect. 6.3.3).

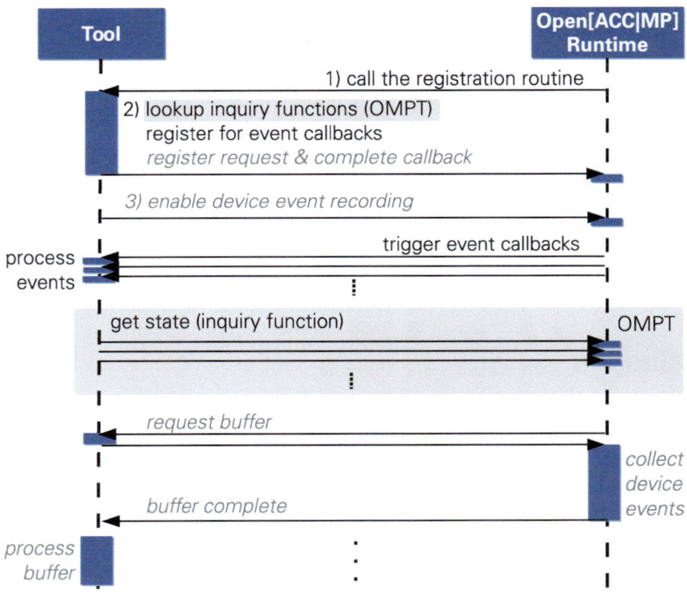

Fig. 6.1 Sequence of interactions between a tool and an OpenACC/OpenMP runtime: For both interfaces the runtime calls the tool initialization routine at first. Then, the tool prepares the data acquisition, e.g. by registering callbacks for events of interest. Sampling of runtime states is only supported by the OMPT interface. Respective activities are highlighted with *grey background color*. Available extension proposals enable portable collection of device data, which is highlighted in *italic grey font*

Both interfaces provide support for instrumentation-based tools using event callbacks. The OMPT interface allows to register event callbacks only within the initialization routine, whereas event registration is allowed at any time during the program execution for ACCT. At application runtime, registered event callbacks are triggered on the host by the OpenMP/OpenACC runtime. The control flow is passed to the performance tool which processes the event.

The OMPT interface also provides support for sampling-based tools using states that have to be maintained by an OpenMP runtime. The tool has to lookup a pointer to the `ompt_get_state` inquiry function during the initialization. The state of the OpenMP runtime can be queried at any time during the program execution.

6.5.2 Tool Integration

Figure 6.2 depicts the data and control flow between application, runtime libraries, the Score-P measurement components and selected analysis tools. For each supported programming model Score-P implements a corresponding component that intercepts the user application, passes the control to the Score-P measurement core

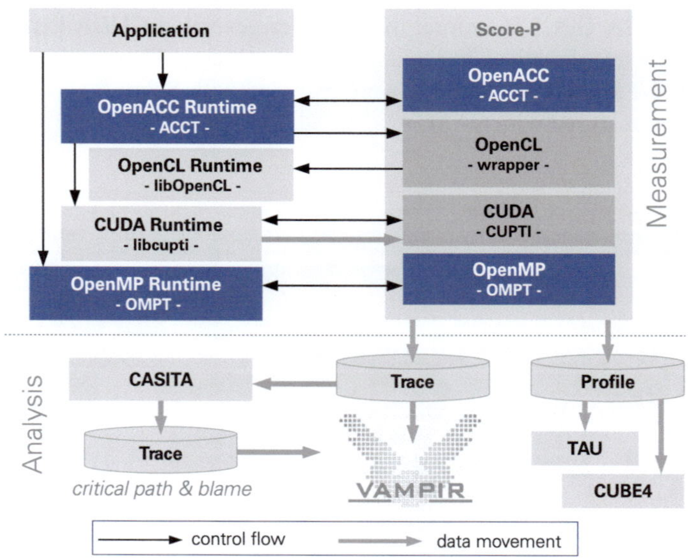

Fig. 6.2 Control and data flow between application, runtime libraries, measurement and analysis components: Assuming an application is using OpenACC and/or OpenMP, Score-P uses the ACCT and OMPT interfaces to collect information on the respective runtime. For low-level programming models such as CUDA and OpenCL it additionally uses CUPTI and an OpenCL wrapper to capture more details. A generated profile can be investigated with TAU or Cube4. A program trace can be directly visualized with Vampir or enhanced by an advanced analysis with CASITA

and captures performance data associated with the current event. After recording performance data the control flow is passed back to the user application, which continues its execution. Figure 6.2 shows the already existing OpenCL adapter which wraps API calls to the OpenCL runtime library and the CUDA adapter using the CUPTI interface to record performance relevant data and events of CUDA applications. To enable event recording of OpenACC and OpenMP applications based on the presented tool interfaces Score-P has been extended by additional adapter components (compare OpenACC and OpenMP boxes in Fig. 6.2). Actually, Score-P uses the OPARI2 source-to-source instrumentation tool to gather performance data of OpenMP programs. OMPT is an alternative approach enabling analysts to look at the OpenMP runtime level that comprises compiler optimization effects, whereas OPARI2 reflects the structure of the source code [10]. The performance data internally collected can be aggregated and written to a profile. Tools like TAU or Cube allow interactive exploration of the profile data and help users to find hot-spots in their application. Alternatively, all individual events can be written to a trace for a detailed analysis of the application's dynamic runtime behavior. For example, the trace file can be visualized with Vampir to interactively investigate dependencies between events happening on different processes. Furthermore, tools such as CASITA implement sophisticated automatic analysis methods for heterogeneous applications, e.g. critical-path-detection, based on trace files. Figure 6.3 shows the visualization of an OpenACC application in Vampir. In this example, the test platform was equipped with an NVIDIA GPU. Therefore, the compiler generated CUDA kernels for the OpenACC *kernels* directives in the source code. OpenACC events on the host as well as CUDA kernels on the GPU were recorded with Score-P and the resulting trace file was analyzed with CASITA.

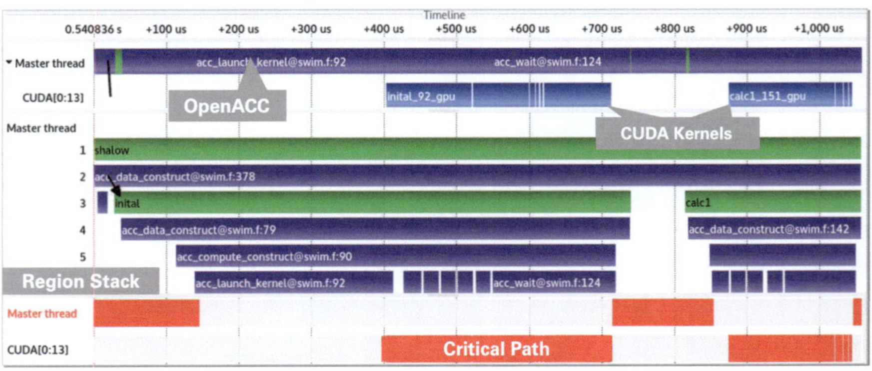

Fig. 6.3 Visualization of an OpenACC application in Vampir: The timeline display on the *top* shows the execution of OpenACC runtime activities on the host CPU. Kernels are launched and a wait operation executed afterwards. CUDA kernels are shown on the *CUDA[0:13]* execution stream. A data transfer between the host CPU and the CUDA device is indicated by a *black line*. The process timeline in the *middle* presents the runtime call stack for all instrumented host activities on the *Master thread*. The *performance radar* timeline on the *bottom* highlights the critical path that has been identified by a trace analysis with CASITA

6.5.3 Comparison of OMPT and ACCT Key Facts

OpenMP and OpenACC provide their features via a set of compiler directives, library routines, and environment variables. The corresponding interfaces OMPT and ACCT aim to provide a standardized way to instrument these compiler directives. From a tool perspective the setup procedure of both interfaces is very similar.

In contrast to ACCT, the OMPT interface provides the possibility to query state information for each OpenMP thread. This feature is necessary for sampling-based performance analysis tools. Both interfaces provide callbacks for event-based performance analysis. Using the ACCT interface a tool can register for up to 25 different callback events. All of them are mandatory and use a common callback signature. The official OMPT specification distinguishes 59 events and 16 runtime states. The revised version defines 67 events and 19 runtime states. However, only a small subset of the events is mandatory and must be supported by all OpenMP runtimes. In addition to mandatory callback events, OMPT specifies optional events which do not have to be implemented by an OpenMP runtime. Furthermore, there are 12 different signatures for callback events defined in the official technical report on OMPT, which has been increased to 23 different signatures in the revised document (Table 6.2).

Implementing the OMPT interface imposes a higher burden on tool developers in comparison to ACCT. For example, tools supporting optional OMPT features have to check their availability and (at best) provide a fallback if specific feature sets are not available on a platform. Both, OMPT and ACCT, provide inquiry functions allowing third-party tools to gather information from the respective runtime. OMPT requires

Table 6.2 OMPT and ACCT key facts according to the official OMPT technical report [5] and the public comment version of the OpenACC 2.5 API [15]

<table>
<tr><th>OMPT</th><th>vs. ACCT</th></tr>
<tr><td colspan="2">Objective:
Analyze the execution of compiler directives</td></tr>
<tr><td colspan="2">Initialization:
Runtime calls tool initialization routine</td></tr>
<tr><td colspan="2">Support for instrumentation-based tools:
Specification of event callbacks (executed on the host) for</td></tr>
<tr><td>59 mandatory or optional events</td><td>25 mandatory events</td></tr>
<tr><td colspan="2">Callback signatures:</td></tr>
<tr><td>12</td><td>1</td></tr>
<tr><td colspan="2">Support for sampling-based tools:</td></tr>
<tr><td>16 mandatory, flexible or optional states</td><td>-</td></tr>
<tr><td colspan="2">Access to tool API:</td></tr>
<tr><td>Lookup routine to obtain pointers to OMPT API inquiry functions</td><td>Directly call ACCT API routines</td></tr>
<tr><td colspan="2">Extension Proposals:</td></tr>
<tr><td>Support for OpenMP 4.0 target devices</td><td>Runtime states for sampling-based tools & Portable device data collection</td></tr>
</table>

the tools to obtain pointers to the individual inquiry functions via a lookup routine. Using the ACCT interface, tools can directly call ACCT API routines.

6.5.4 Extension Proposals

As the OpenMP and OpenACC standards continue to evolve the corresponding tool interface specifications need to keep up with the changes. The official OMPT interface, as described in the current technical report [5], provides support only for OpenMP 3.1 features. For instance target constructs which were introduced in OpenMP 4.0 are not yet considered. However, an OMPT extension proposal for target devices has been developed by Cramer et al. in [1]. Here, additional runtime begin/end event pairs for *target*, *target data* and *target update* regions including the corresponding type signatures are defined. Furthermore, events for synchronous or asynchronous data mapping (e.g. data transfers) for each variable were added. The strength of this approach is that the internal runtime information can be used in order to determine fine granular data mapping/transfer times to or from a target device (e.g. a GPGPU). In addition, the extension proposal defines four new inquiry functions. Two of them allow to get information about the current device ID or target region ID. This is necessary, because calling the corresponding OpenMP runtime library routine within a callback is unsafe and might cause a deadlock. Due to the fact that target devices may use a different clock generator than the host device the proposal intends to use an inquiry function in order to determine the actual device time which allows a tool to bring host and device events into a certain temporal order. The last proposed inquiry function reflects the fact that it may not be practical or possible to trigger event callbacks on the host for events occurring on a target device. Hence, a mechanism to gather events on a target device and transfer them to the host is required.

Based on this proposal the official technical report [5] was revised [7]. Concerning the runtime events, the main change of the proposal in [1] reflects the latest developments for the upcoming OpenMP 4.5 specification, especially the definition of a *target task*, which allows asynchronous data mappings/transfers. The inquiry function for the transfer of the collected target events was replaced by an asynchronous buffering API. In this approach the OpenMP runtime interrupts the execution on the host by invoking a corresponding callback function when a new trace buffer is required. The memory has to be provided by the tool and is filled by the device runtime. When the buffer is full or the application exits, the device provides it to the OpenMP runtime on the host that triggers a respective callback in the tool. Furthermore, the revised technical report adds an interface for collecting device specific (native) events such as those generated by CUPTI [12].

Sampling-based tools are currently not supported by the ACCT interface, because it neither defines OpenACC runtime states nor a respective query routine. The third argument of the tool initialization routine `acc_register_library` is preserved to add support for sampling runtime states in future specifications. An extension proposal to cover sampling with the ACCT interface has been published in [4]. It describes a state query routine `acc_prof_get_state` that collects the state, the active device and the activity queue a querying thread is waiting for. Three fundamental states (`wait_data`, `wait_compute` and `wait_acc`) are defined. Optionally, an OpenACC runtime might provide a more detailed version of these states, e.g. to distinguish waiting in kernels or parallel regions, waiting for data allocation, data upload or data download. The extension proposal clearly addresses the host-directed execution model with asynchronous accelerator activities in OpenACC and can be used to identify wait states and assign the waiting costs to respective causes.

The same publication also proposes to extend the ACCT interface for portable device data collection. The proposal is similar to the concept that is used in the CUPTI activity API [12] and the asynchronous buffering API that has been introduced in the revised OMPT specification. Figure 6.1 illustrates the concept (activities in italic grey font). During the execution of the initialization routine the tool registers a buffer request and a buffer complete callback. Before a runtime starts to record events on the device the tool has to enable recording. If recording is enabled, the runtime will request a buffer before the device gets active and trigger a buffer complete callback when all records written to the buffer are valid. The tool can then process completed buffers. The proposed extension concept is reasonable and has been proven to work with the CUPTI activity API. As CUDA is probably the most often used target for OpenACC applications the implementation effort for OpenACC runtime developers might not be too costly. Otherwise, it might be sufficient to use CUPTI for data collection on CUDA devices or other approaches for measuring the target API as for example described in [2] for OpenCL.

The runtime overhead induced by the presented tools interfaces has been investigated in [10] for OMPT and in [4] for ACCT. For both, the measurement overhead is low in typical usage scenarios of OpenMP and OpenACC. As with other instrumentation-based approaches the measurement of extremely short-running code regions is expensive in terms of additional runtime. Therefore, programs with inefficient usage of OpenMP or OpenACC, such as many extremely short running kernels or tiny tasks, result in a notable performance degradation.

6.6 Conclusion

We discussed performance analysis for OpenACC and OpenMP based on their most recent tool interfaces and respective extension proposals. Both interfaces provide tool developers with a portable approach to obtain performance relevant information on programs utilizing OpenACC and/or OpenMP. We compared their design and highlighted similarities as well as differences. Based on the implementation in the Score-P analysis infrastructure we evaluated the benefit and the applicability of both interfaces for instrumentation-based performance analysis. The Vampir visualization of an execution trace that has been enhanced with CASITA shows the level of detail performance analysis can provide using the presented interfaces.

We illustrated drawbacks in the current versions of OMPT and ACCT and considered recent extension proposals that address missing functionality. Despite the fact that both, the OMPT and the ACCT interface, are not in their final states yet, these tool interfaces are extremely valuable for the design of portable and robust performance analysis tools.

References

1. Cramer, T., Dietrich, R., Terboven, C., Müller, M.S., Nagel, W.E.: Performance analysis for target devices with the OpenMP tools interface. In: IEEE International Parallel and Distributed Processing Symposium Workshop (IPDPSW), pp. 215–224. IEEE (2015)
2. Dietrich, R., Tschüter, R.: A generic infrastructure for OpenCL performance analysis. In: 8th International Conference on Intelligent Data Acquisition and Advanced Computing Systems: Technology and Applications (IDAACS). IEEE (2015)
3. Dietrich, R., Schmitt, F., Grund, A., Schmidl, D.: Performance measurement for the OpenMP 4.0 offloading model. Euro-Par 2014: Parallel Processing Workshops. Lecture Notes in Computer Science, vol. 8806, pp. 291–301. Springer International Publishing, Cham (2014)
4. Dietrich, R., Juckeland, G., Wolfe, M.: OpenACC programs examined: a performance analysis approach. In: 44th International Conference on Parallel Processing (ICPP). IEEE (2015)
5. Eichenberger, A., Mellor-Crummey, J., Schulz, M., Copty, N., Cownie, J., Dietrich, R., Liu, X., Loh, E., Lorenz, D., et al.: OpenMP Tools Working Group: OpenMP Technical Report 2 on the OMPT Interface. The OpenMP Architecture Review Board (2014). http://openmp.org/mp-documents/ompt-tr2.pdf, 23 October 2015
6. Eichenberger, A.E., Mellor-Crummey, J., Schulz, M., Wong, M., Copty, N., Dietrich, R., Liu, X., Loh, E., Lorenz, D.: OMPT: an OpenMP tools application programming interface for performance analysis. OpenMP in the Era of Low Power Devices and Accelerators. Lecture Notes in Computer Science, vol. 8122, pp. 171–185. Springer, Berlin (2013)
7. Eichenberger, A.E., Mellor-Crummey, J., Schulz, M., Copty, N., Cownie, J., Cramer, T., Dietrich, R., Liu, X., Loh, E., Lorenz, D.: OMPT: An OpenMP Tools Application Programming Interface for Performance Analysis. Revised 5 October 2015. https://github.com/OpenMPToolsInterface/OMPT-Technical-Report, 19 October 2015
8. Itzkowitz, M., Mazurov, O., Copty, N., Lin, Y.: An OpenMP runtime API for profiling. OpenMP ARB as an official ARB white paper available online at http://www.compunity.org/futures/omp-api.html, vol. 314, pp. 181–190 (2007)
9. Liao, C., Quinlan, D.J., Panas, T., de Supinski, B.R.: A ROSE-based OpenMP 3.0 research compiler supporting multiple runtime libraries. Beyond Loop Level Parallelism in OpenMP: Accelerators, Tasking and More, pp. 15–28. Springer, New York (2010)
10. Lorenz, D., Dietrich, R., Tschüter, R., Wolf, F.: A comparison between OPARI2 and the OpenMP tools interface in the context of Score-P. In: Proceedings of the 10th International Workshop on OpenMP (IWOMP), Salvador, Brazil. LNCS, vol. 8766, pp. 161–172. Springer International Publishing (2014)
11. Mohr, B., Malony, A., Hoppe, H.C., Schlimbach, F., Haab, G., Shah, S.: A performance monitoring interface for OpenMP. In: Proceedings of the 4th European Workshop on OpenMP (EWOMP'02), Rom, Italien (2002)
12. NVIDIA: CUDA Toolkit Documentation – CUPTI (2014). http://docs.nvidia.com/cuda/cupti/index.html, 12 June 2015
13. OpenACC-Standard.org: Interfacing Profile and Trace tools with OpenACC Programs. Technical report, OpenACC-Standard.org (2014). http://www.openacc.org/sites/default/files/TR-14-2_0.pdf, 12 June 2015. Technical Report TR-14-2

14. OpenMP Application Program Interface, Version 4.0 (2013). http://www.openmp.org/mp-documents/OpenMP4.0.0.pdf, 16 October 2015
15. The OpenACC Application Programming Interface, Version 2.5, Public Comment Version (2015). http://www.openacc.org–Specification&TechReports, 14 October 2015

Chapter 7
Extending MUST to Check Hybrid-Parallel Programs for Correctness Using the OpenMP Tools Interface

Tim Cramer, Felix Münchhalfen, Christian Terboven, Tobias Hilbrich and Matthias S. Müller

Abstract Current High Performance Computing (HPC) systems consist of compute nodes that can communicate via an interconnect. Each compute node features multiple compute cores that can access shared-memory. The Message Passing Interface (MPI) is the de-facto standard for the programming of distributed memory applications. At the same time, OpenMP is a well-suited parallel programming paradigm to utilize the parallel cores within a compute node. Thus, current HPC systems encourage a hybrid programming approach that combines MPI with OpenMP. However, using both programming paradigms at the same time can lead to more error-prone applications. The runtime correctness checking tool MUST supports programmers in the detection and removal of MPI-specific programming defects. We present an extension of MUST towards the analysis of OpenMP-MPI parallel applications in order to support programmers that combine both paradigms. This includes thread-safety concerns in MUST itself, an extended event model based on the upcoming OpenMP Tools Interface (OMPT), as well as a prototypical error analysis with a synthetic example. We further discuss classes of defects that are specific to OpenMP applications and highlight techniques for their detection.

T. Cramer (✉) · F. Münchhalfen · C. Terboven · M.S. Müller
IT Center, RWTH Aachen University, 52074 Aachen, Germany
e-mail: cramer@itc.rwth-aachen.de

T. Cramer · F. Münchhalfen · C. Terboven · M.S. Müller
Chair for High Performance Computing, RWTH Aachen University, 52074 Aachen, Germany

T. Cramer · F. Münchhalfen · C. Terboven · M.S. Müller
JARA - High-Performance Computing, Schinkelstraße 2, 52062 Aachen, Germany

F. Münchhalfen
e-mail: munchhalfen@itc.rwth-aachen.de

C. Terboven
e-mail: terboven@itc.rwth-aachen.de

M.S. Müller
e-mail: muller@itc.rwth-aachen.de

T. Hilbrich
Technische Universität Dresden, 01062 Dresden, Germany

A. Knüpfer et al. (eds.), *Tools for High Performance Computing 2015*,
DOI 10.1007/978-3-319-39589-0_7

7.1 Introduction

Current high performance computing systems are based on highly parallel architectures. Usually, they are clustered shared memory systems, which motivates a combined use of a message passing paradigm and a shared memory programming paradigm, in order to develop good performing and scalable applications. The Message Passing Interface (MPI) [9] is still the de-facto standard for the programming of distributed memory applications and OpenMP [11] for the programming of shared memory applications. Unfortunately, parallel programming itself is complex and error-prone—independent of the programming paradigm. Combining shared memory paradigms like OpenMP with MPI increases the applications complexity additionally. The diversity of these error classes is comprised of the class of "traditional" OpenMP usage errors, like race conditions, the class of "traditional" MPI usage errors, like deadlocks, and a combinatorial class that results from the mix of the paradigms.

While many program defects can be corrected with the help of debuggers or static code analysis techniques, removing some defects is highly inefficient with these approaches. Runtime error detection tools combine automatic error detection capabilities with the capability to analyze realistic workloads. The MUST [4] tool is one such runtime tool that detects wide ranges of MPI usage errors. MUST not only reports information on the presence of errors, but also provides state information on the origin of these errors, often indicating root causes directly. Particularly, MUST can detect errors that are not (yet) visible. Correcting these kind of errors increases program portability across different systems or software stacks, which is especially important for codes that are widely used.

We present extensions to the infrastructure of MUST that enable correctness checking for combined OpenMP-MPI programming. This includes work on correctness analyses in MUST itself, as well as work on its underlying tools infrastructure GTI [5], such as new communication services and thread safety in the base services PnMPI [13]. Furthermore, we will analyze the feasibility of OpenMP specific instrumentation techniques with a focus on a recent proposal for a standardized OpenMP Tools Interface (OMPT) [3].

The remainder of this paper is structured as follows. Section 7.2 presents an OpenMP error classification, which allows an analysis of the usability of OMPT for runtime correctness checking (Sect. 7.3). In Sect. 7.4 we present our modifications and extensions of the MUST infrastructure. Finally, Sect. 7.5 evaluates our modifications for applicability.

7.2 OpenMP Error Classification

Existing classifications for OpenMP usage errors provide a summary of typical error situations [10, 12, 14]. We follow a recent classification [10] that assigns OpenMP programming errors to individual error classes and identifies techniques for automatic detection. This classification solely focuses on OpenMP 3.0 features and the offloading features introduced with OpenMP 4.0 [11]. It provides a framework for common OpenMP programming defects and can be extended in the future, if further error classes arise with newer OpenMP standards.

Version 3.0 of the OpenMP standard adds *explicit tasks*, which establish a paradigm extension in OpenMP. The original OpenMP fork-join model utilizes worksharing constructs that often require a-priori workload knowledge for efficient parallelization. Tasking provides a natural paradigm to support more dynamic workloads, such as graph traversal. In version 4.0, the OpenMP standard additionally distinguishes between *host* and *target* devices. The two terms correspond to a host computer and accelerator targets respectively. Code that is enclosed in a `target` region is then compiled for execution on the target accelerator. This extension enables OpenMP to utilize heterogeneous architectures that previously required additional programming models like CUDA, OpenCL, or OpenACC.

The classification in Fig. 7.1 follows our previous study [10] with an incorporation of existing studies [12, 14]. Particularly, we incorporate all types of defects of which we are aware, as well as novel defects that the OpenMP 3.0 and OpenMP 4.0 extensions introduce.

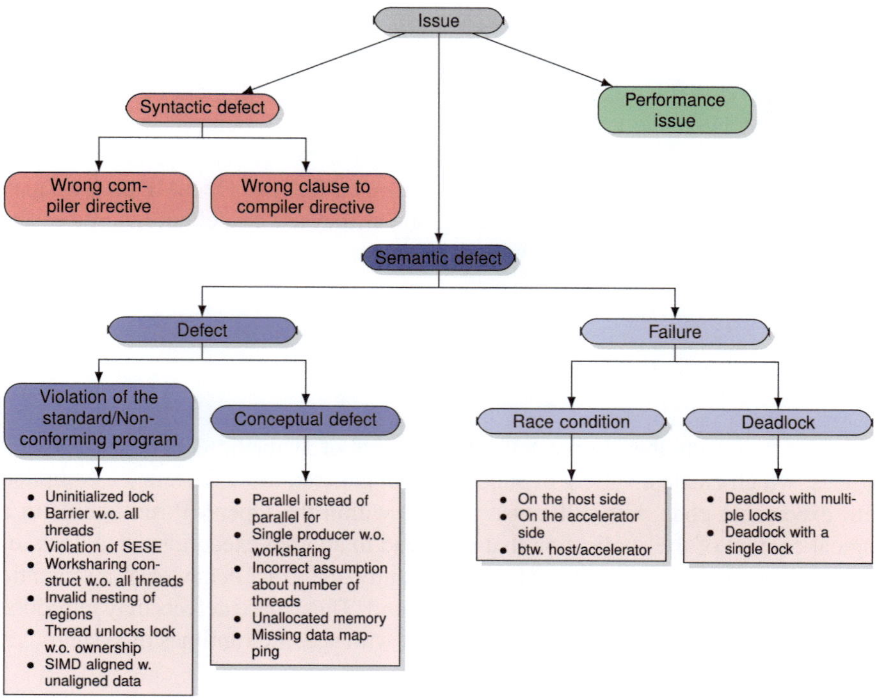

Fig. 7.1 Classification of common issues in OpenMP applications [10]

The naming *error* is ambiguous and may refer to a programming mistake, to the way the mistake manifests at execution time, or to an incorrect result of an application. Therefore, we use the following nomenclature [6, 15]:

- *defect* to address incorrect source code; and
- *failure* visible manifestation of a defect, e.g., aborted execution, incorrect result, or deadlock.

We distinguish defects from failures since automatic tools will often only spot a failure. Failure often coincide with their underlying defects, but for some cases a runtime tool detects a failure only well after the execution of the underlying defect. Thus, in these situations a trace-back to the defect—in the form of a source code location, e.g., line and file—is necessary. Correctness tools usually detect either defects or failures. A compiler or static analysis tool for example will typically spot defects directly. Runtime analysis tools and debuggers may also do so, but often they will only highlight a resulting failure instead. We include this notion in our classification to highlight our expectation towards how the defect would become visible in practice.

7.2.1 Syntactic Defects

We classify code defects that violate compliance with the grammar of the underlying programming language as *syntactic defects*. For OpenMP defects, we can restrict this class to compiler directives, and thus, distinguish between mistyped and correct OpenMP prefixes (e.g. `#pragma omp` in C/C++). Mistyped OpenMP prefixes may influence the correctness of a parallel program, as they are by-default ignored by the compiler (while it most probably will give out warnings).

7.2.2 Semantic Defects

We classify defects that do not fall into the class of syntactic defects as *semantic defects*, which create a variety of subclasses. In general, these defects are compiled into executable code, but will cause failures within the OpenMP runtime or in a logical context of the application that may lead to aborted executions, deadlock, or incorrect/irreproducible results. Most importantly, the exact behavior of a semantic defect depends on the runtime and system at hand. Thus, these defects may introduce portability problems that only become visible with specific runtimes or systems.

7.2.3 Detectability

Table 7.1 evaluates each of the error classes in Fig. 7.1 in terms of automatic detectability by a tool. We distinguish between typical tool approaches: Compilers, compiler-based static analysis (including inter-procedural analysis, symbolic execution, and model checking), the OpenMP runtime, debuggers, and runtime tools. As the table shows, the type of tool must be chosen depending on the type of defect (or failure) that has to be detected, as no single type of correctness checking tool is able to detect all error classes.

Table 7.1 Defects from Fig. 7.1 and their detectability: no mark indicates that we expect the type of tool to not be able to detect the defect; (•) indicates that we expect the type of tool to be ab able to detect the defect, but that it would typically not implement this functionality; • indicates that we expect the tool to be able to detect the defect and that it would typically implements such functionality [10]

#	Mistake	Compiler	Static analysis	OpenMP runtime	Debuggers	Runtime tools
	Syntactic mistakes					
1.	wrong_directive		•			
2.	wrong_clause	•	•			
	Semantic mistakes					
	Violation of the standard					
3.	uninitialized_locks			•	•	•
4.	barrier_wo_all_threads		•	•		•
5.	violation_sese	(•)	•			
6.	worksharing_wo_all_threads					
7.	invalid_nesting	(•)	•	•		(•)
8.	lock_unlock_nonowner		•	•		•
9.	simd_aligned		(•)		•	
	Conceptual defect					
10.	parallel_inst_parallel_for					
11.	single_prod_wo_worksharing					
12.	number_of_threads					
13.	unallocated_memory	(•)	•	•	•	•
14.	missing_data_mapping	(•)	•			(•)
	Race condition					
15.	host			(•)		•
16.	accelerator			(•)		•
17.	host_accelerator					(•)
	Deadlock					
18.	multiple_locks		(•)			•
19.	single_lock		(•)			•

7.3 OpenMP Tools Interface

The OpenMP architecture review board has recently published an Application Programming Interface (API) for first-party performance tools as a technical report [2]. This OpenMP tools interface (OMPT) describes a standardized and portable API to construct powerful performance analysis tools as a possible future extension to the OpenMP specification. In comparison to source-to-source instrumentation, this approach enables a tool to gather higher-level detail information about the application and the OpenMP runtime system, with low overhead. Furthermore, OMPT does not only provide support for event-based performance tools, but also for sampling-based performance tools by defining and maintaining states of the OpenMP runtime. The runtime should associate the activity of each thread to a state at any point of time, which allows a tool to analyze the behavior of an application. An important design objective of OMPT is that adding the API should not impose unreasonable implementation burdens, on both the runtime and the tool developer. Thus, increasing chances for acceptance and standardization. For this reason, the interface defines a small set of mandatory features (events and states), and extends it by an additional set of optional features. This enables a performance tool to rely on functionality common to all OpenMP runtimes, as well as a possibility to gather and analyze more detailed information about program behavior at runtime.

Although the interface was designed for performance analysis, we highlight that it suffices for use in correctness checking tools as well, including an identification of limitations. With respect to the error classification in Sect. 7.2, we concentrate on semantic defects, since syntactic mistakes are poorly detectable with the help of an instrumented runtime system. Note that an identification of performance defects is not the scope of this work, and is considered by other approaches already [1, 7, 8]. In the following, we discuss the usefulness of the OpenMP tools interface for some individual defect classes (refer to Table 7.1).

Uninitialized locks: The OpenMP standard demands that each regular (or nested) lock has to be initialized before it is acquired, the behavior is unspecified otherwise (refer to #3 in Table 7.1). The following OMPT events have to be collected by a correctness checking tool in order to find this error:

- `ompt_event_init_lock`
- `ompt_event_acquired_lock`
- `ompt_event_acquired_nest_lock_first`
- `ompt_event_acquired_nest_lock_next`

All of these events provide a unique wait identifier that can be used to distinguish between different locks. To detect this kind of defect, MUST manages corresponding event lists and will report an issue as soon as an acquire event was triggered, before an init event with the same wait identifier (the same lock), or if the init event is missing. However, one has to keep in mind that runtime behavior for such a situation is not well defined and implementations might behave differently. For instance, the current Intel OpenMP runtime (version 16.0) aborts execution with a segmentation violation while the GNU runtime (version 4.9) executes the example without any warnings or error messages. For both runtime implementations the event will be triggered, so that even if the runtime crashes, the defect can be detected. However, a correctness checking tool cannot guarantee to detect such defects under all circumstances, either because of yet unknown side effects or a unexpected runtime crash before the corresponding events could be received.

Barrier not reached by all threads of a team: Although synchronization barriers are less error prone than locks, they offer potential for a violation of the "Single Entry, Single Exit" (SESE) principle (refer to #4 and #5 in Table 7.1). This is the case, if a barrier is not reached by all thread of the same team or the threads of a team reach different barriers. Listing 7.1 shows an hybrid example code for such a violation. Depending on runtime behavior, the defect can manifest as deadlock, aborted execution, or may remain unnoticed. In the deadlock case, threads would wait for each other to reach their respective barrier. Both the Intel and GNU compilers neither report an error during compilation nor during execution time. The two different code paths are just executed alternating by all threads and the two different barriers are incorrectly merged towards a single synchronization, so that the application terminates without failure. In case of low code complexity, such kinds of errors are detectable with static code analysis. However, in this example, behavior is dynamic, due to the function call in the `if` statement. Runtime analysis can follow an actual execution in order to analyze such dynamic behavior. For the detection with OMPT, we use the events `ompt_event_barrier_begin` and `ompt_event_barrier_end`. However, the interface does not provide unique barrier identifiers. To overcome this limitation we use libunwind.[1] We determine the source code location of the barrier by resolving its first return address on the call stack and check whether it is not within the OpenMP runtime.

```
MPI_Init(...);
#pragma omp parallel
{
    if( omp_get_thread_num() % 2 ){
        #pragma omp barrier
        printf("Barrier 1\n");
    }else{
        #pragma omp barrier
        printf("Barrier 2\n");
    }
}
MPI_Finalize();
```

Listing 7.1 A hard to detect defect that fails to let all threads of a team reach the same OpenMP barrier construct

Invalid nesting: The OpenMP standard strictly forbids programmers to closely nest worksharing constructs inside each other, because each worksharing requires the context of a single parallel region (refer to #7 in Table 7.1). Although the compiler is able to identify this type of defect in many cases, there are some situations in which it may fail to detect them. These situations are orphaned OpenMP regions and regions

[1] http://www.nongnu.org/libunwind.

inside third-party libraries. Listings 7.2 and 7.3 show an example for such an invalid nesting. Within an OpenMP loop construct, the function foo is called, which contains another loop construct. For all OpenMP worksharing constructs (loop, `sections`, `single` and `workshare`) corresponding OMPT begin and end events exist:

- `ompt_event_loop_{begin,end}`
- `ompt_event_section_{begin,end}`
- `ompt_event_singe_in_{begin,end}`
- `ompt_event_workshare_{begin,end}`

Since these events all provide the same signature with the `parallel_id` of the encountered parallel region, a tool can detect nested worksharing constructs with the same `parallel_id` and report the defect.

```
int main(){
    #pragma omp parallel
    {
    #pragma omp for
    for(int i=0; i < N; i++)
        foo();
    }
    return 0;
}
```

Listing 7.2 Invalid nesting (main.c)

```
void foo(){
    #pragma omp for
    for(int i=0; i < N; i++)
        printf("I am a thread\n");
}
```

Listing 7.3 Invalid nesting (foo.c)

Use of `parallel` instead of `parallel for`: An example for a conceptual defect is the unintended use of a syntactically correct, but semantically misapplied directive (refer to #10 in Table 7.1). The initialization routine in Listing 7.4 uses a `parallel` directive instead of a `parallel for` directive. Thus, instead of a parallel initialization where each thread handles a part of the array, all threads do exactly the same work and initialize the complete array. Since the computational result is still the same as for the correct execution, the user will not recognize this defect directly, although the parallelization will not yield any performance benefit. This highlights a particular limitation of OMPT, as well as a more general limitation for runtime correctness approaches. The actual defect cannot be detected with OMPT, since on the one hand the intention of the user is unknown, and on the other hand OMPT delivers not enough information (e.g., about the actual accessed memory address). However, the defect manifests as a race condition that can be detected with other techniques. One possible method is to trace all memory accesses during runtime, which can e.g., be done with the help of binary instrumentation. The downside of this is that such a memory tracing is very expensive in terms of runtime and memory overheads. In order to enable MUST to deal with this kind of issues in the future, a corresponding component is under development.

```
void init(int* a) {

    #pragma omp parallel
    {
        for (int i=0; i < N; i++)
            a[i] = 42;
    }

    return;
}
```

Listing 7.4 Conceptual defect: Semantically wrong directive

```
int sum = 0;
// no copy back of sum after
// execution of target region
#pragma omp target map(to:sum)
{
    for (int i=0; i < N; i++)
        sum += i;
}
// read outdated sum value on host
printf("%d", sum);
```

Listing 7.5 Conceptual defect: Missing data mapping

Missing data mapping from target device: Another example for a conceptual defect is the use of outdated data on a host device (refer to #14 in Table 7.1). In Listing 7.5 a scalar value `sum` is mapped from a host device to a target device, e.g., a GPGPU device. The target then modifies the data, but does not map it back. Thus, on the host, the value becomes outdated. With the help of a memory tracer in combination with OMPT, MUST can report an error when an address is modified on a target device, but not mapped back, although the result is accessed on the host. The behavior of the given example is undefined according to the OpenMP standard, even if it was the intention of the programmer to omit the mapping back to the host device. Thus, with one implementation the host could work on the old data, while on another it may operate on the new data.

Race condition between host and target device: OpenMP `target` constructs create a new class of race conditions between host and target devices (refer to #17 in Table 7.1). In Listing 7.6 one part of the `result` is computed on the host and another part on the target device. However, the result is not deterministic, because it depends on the order in which the individual invocations of `compute()` terminate. If the target device executes the routine faster then the host the result is correct. Otherwise, the partial result of the host will be overwritten as soon as the result is mapped back from the device.

Since the current technical report published by the OpenMP Architecture Review Board (ARB) [2] only covers OpenMP 3.1, we proposed an extension for `target` constructs in [1]. Parts of this proposal have already been discussed in the OpenMP language committee and are already reflected in the revised technical report [3]. This includes an extended signature for the event `ompt_event_target_data_map_begin` where a host and a device address is delivered. These addresses in combination with a memory tracer can be used to detect the new type of race conditions.

```
double result = 0;
#pragma omp parallel num_threads(2)
{
    #pragma omp sections
    {
        #pragma omp section
        #pragma omp target map(tofrom:result)
        {
            result += compute();
        }
        #pragma omp section
        {
            result += compute();
        }
    }
}
```

Listing 7.6 Race condition between host and target device

7.4 MUST Extensions

To enable the potential of correctness checks for hybrid applications in MUST, several changes to the integral components on which the correctness checking software builds had to be incorporated. These include changes to the mechanism that is used to intercept library calls, as well as changes to the internal communication strategies and layout. Changes which were necessary to ensure thread-safety in MUST are described in Sect. 7.4.1. Section 7.4.2 then details changes to the communication infrastructure of MUST, as well as consequences for event handling.

7.4.1 Thread-Safety

A runtime error detection tool analyzing applications that mix MPI and OpenMP has to meet thread-safety requirements itself. In the case of MUST [4] this includes adaptations to its underlying components GTI [5] and PnMPI [13].

PnMPI is the lowest level in the MUST software stack and manages MPI wrapping services. Since MPI operations can be issued by multiple threads of a process, this service forms a first target for thread-safety extensions. Historically, PMPI tools overwrite some symbols of the MPI implementation, e.g., to intercept `MPI_Send` they would include a symbol of similar name themselves. Thus, an application call to an MPI function would resolve to the tool, instead of to the MPI library. In order to let the tool forward intercepted calls to the actual MPI implementation, a second symbol for each MPI function (e.g. `PMPI_Send` for `MPI_Send`) exists. Both symbols then point to the same routine in the MPI library. PnMPI uses this interception mechanism, but extents it such that multiple tools, or so called *modules*, can intercept MPI functions at the same time. The PnMPI library overwrites all MPI symbols to intercept its function calls itself. The library then loads in additional modules that want to intercept MPI functions. PnMPI modifies the libraries of these modules with a symbol rewriting technique that replaces symbols prefixed with *PMPI* with *XMPI*. For each MPI function, PnMPI provides a function with prefix *XMPI*. Thus, when a module loaded by PnMPI issues a PMPI call, it calls into PnMPI instead of the MPI library. This allows PnMPI to create a recursion that calls functions of the same name (e.g. `MPI_Send`) into different modules. Figure 7.2 shows how PnMPI recursively traverses different wrapper functions for MPI functions in different PnMPI modules. Each time a wrapper function makes it's call to the PMPI implementation, which was replaced by XMPI as mentioned, the flow of execution goes back into the PnMPI library. The PnMPI library will then successively call wrapper implementations of that MPI function in the remaining loaded modules (refer to Fig. 7.2).

The recursion in PnMPI uses control structures to store information such as the recursion level. In a multi-threaded application, our first extension is to move these structures into Thread Local Storage (TLS). Thus, multiple threads are able to traverse the recursion mechanism simultaneously, without interfering with the recursion steps of the other threads. Since we are using TLS to separate this thread local information, we do not leverage any kind of synchronization at this stage. Without the thread-safety modifications we implemented, the mechanism would only work if one thread at a time would access MPI functions, e.g., as in the case of *MPI_THREAD_SINGLE*.

Besides thread-safety measures for PnMPI, further modifications in the internal communication layout of GTI and MUST allow us to support hybrid applications. MUST uses both the application processes/threads for its correctness analysis, as well as further processes—*tool processes*—that allow an offloading of complex analyses. MUST then configures a communication layout to connect application processes with the tool processes. To analyze hybrid applications, we use a new type of layout.

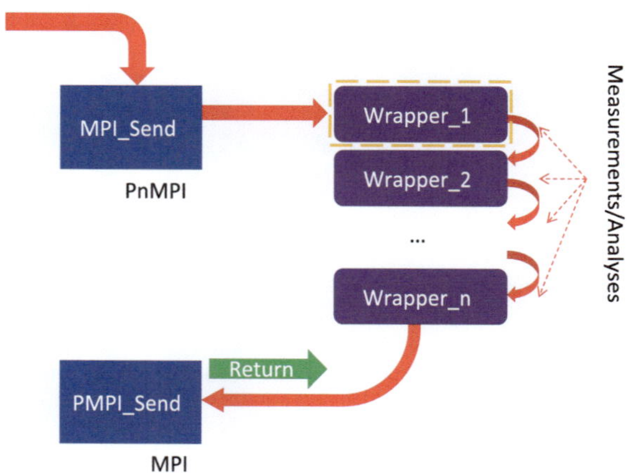

Fig. 7.2 PnMPI calls one PMPI wrapper module after another and finally gives control to the real MPI function

With the presence of threads, MUST has to differentiate correctness analyses that refer to a single thread, a single process, or non-local analyses that bring information of multiple processes together. Originally, it assumed a single thread per process only. So MUST did not have to discern correctness analyses for threads from correctness analyses for processes. With the presence of multiple threads, we cannot execute correctness analyses that use state information for a process on a single application thread. Instead we must combine information from all threads of a process to run such analyses. We use a new layout that employs an additional tool-owned thread—*tool thread*—for that purpose. Tool threads receive information from all threads of their associated process, as to run correctness analyses that refer to the overall state of a process. Tool threads then forward information towards MUST's tool processes. Thus, tool threads forward information from individual application threads sequentially onward. This property helps to maintain an event forwarding that retains basic event order between the threads. As an example, an event that occurs before a thread joins an OpenMP barrier cannot overtake an event that another thread creates after the barrier.

Figure 7.3 illustrates the thread-to-thread communication system. Tool threads and tool processes can communicate with a variety of communication mechanisms in GTI, including TCP/IP, Linux message queues, and MPI communication. The application threads communicate with their associated tool thread using shared message queues. The receiver side then processes all of the thread message queues sequentially as to retrain event order between threads. Each thread has its own queues for reading and writing, so that synchronization only occurs with the tool-thread. This way, application threads do not forcibly synchronize when GTI events occur.

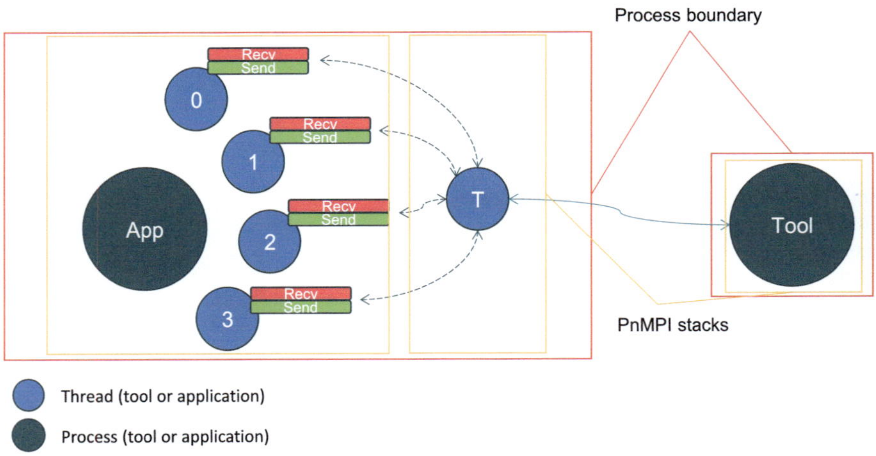

Fig. 7.3 Illustration of the communication between application threads and their tool thread, as well as the communication of a tool thread with MUST's tool processes

7.4.2 Extended Event Model

The existing MPI correctness analysis of MUST requires information on the invocation of MPI operations. The interrogation technique with wrapper functions (previous section) provides this information. As to facilitate efficient creation of these wrappers, MUST incorporates a generator that maps (correctness) analyses to their target MPI operations. This also allows user driven selection of correctness analyses and offloading capabilities, which influence tool overhead. We extend this generation mechanism to support the callback functions that OMPT uses, including their automatic registration. The main difference between MPI wrapper functions and OMPT callbacks is that MPI wrapper functions must invoke the MPI implementation in turn. Thus, correctness analyses for MPI can occur before, called *PRE* order, or after, called *POST* order, the invocation of the MPI library. OMPT callbacks are event notifications only and do not need to call into the OpenMP runtime. Consequently we omit the distinction of analysis-order for these events.

An implementation of OMPT is available as part of a modified version of the Intel/LLVM OpenMP runtime.[2] This runtime exports the weak symbol *ompt_initialize* that a tool can overload. Thus, the dynamic loader (or in case of static linking the linker) resolves this symbol, which also requires that the overloading happens when shared libraries are resolved. The *ompt_initialize* function then provides handles to OMPT functions to the target tool. The OpenMP runtime calls it as soon as the application starts up, which is usually shortly after the program entry point. The function pointers that *ompt_initialize* provides allow a target tool to resolve other relevant OMPT functions. This includes registration of callback handlers and queries to internal information of the runtime.

Since MUST is a collection of PnMPI modules, i.e., dynamically loadable libraries, none of MUST's functions is visible to the dynamic loader. PnMPI is responsible for loading and calling MUST modules and it does so using *dlopen* and *dlsym* at runtime. Consequentially, we had to incorporate *ompt_initialize* as an additional wrapper function into PnMPI, as to make it visible to the dynamic loader already. Thus, PnMPI is able to distribute the occurrence of *ompt_initialize* to any MUST modules that requires it. We use the recursion technique from PnMPI too allow multiple tools to observe *ompt_initialize*. A new MUST module receives this information and stores the function pointers that it provides internally. Every MUST analysis that uses OMPT can specify this module as an dependency and use it as an interface to register OMPT callbacks, or to get internal information of the runtime.

Figure 7.4 depicts the MUST software stack and key modifications that we incorporate to support MPI-OpenMP applications. In the existing components, PnMPI provides MPI wrappers and forwards them to MUST modules that require them. GTI handles the generation of the wrappers that the MUST modules use. A new GTI component registers OMPT callbacks with an OpenMP runtime that supports OMPT. A MUST module then receives this information and uses an epoch model to

[2] http://openmp.llvm.org.

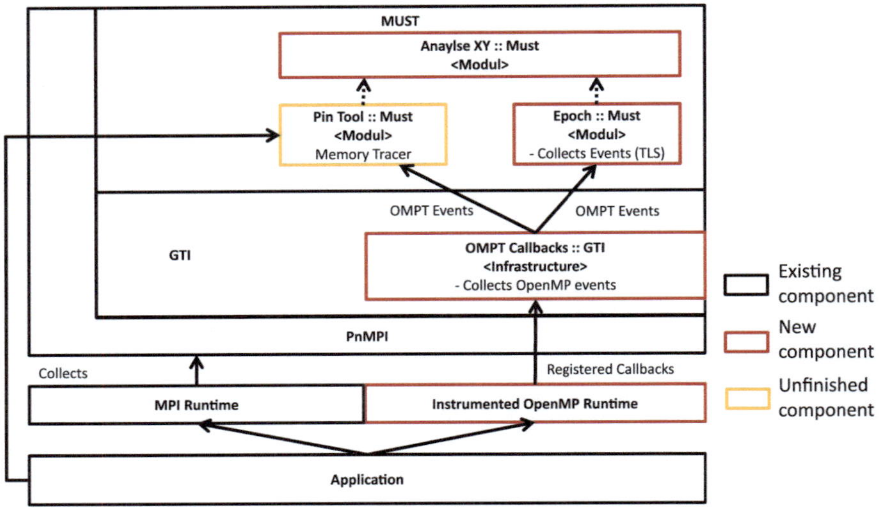

Fig. 7.4 Architectural diagram of the MUST software stack highlights modifications for MPI-OpenMP applications

evaluate it for correctness. Furthermore, we envision a memory tracing component that uses binary instrumentation to provide MUST information on read and write accesses. Since memory tracing is expensive in terms of runtime and memory consumption, we plan to use information about OMPT events to steer compression of this access information. We can compress any memory accesses between two synchronization events, e.g., by detecting contiguous or strided accesses. As we illustrate in Sect. 7.3, this combined information supports the detection of complex error classes in MPI-OpenMP applications.

7.5 MUST Evaluation Run

To test the applicability and soundness of our approach we applied our extended prototype to the example code in Listing 7.1. On each MPI process, the example uses distinct threads of a single team to issue distinct OpenMP barriers concurrently. This violates the OpenMP standard and renders the application unsafe. Due to the lack of unique barrier identifiers for the barrier events of OMPT, we use libunwind to determine their source code location for distinction. We then apply the correctness check that we illustrated in Sect. 7.3. Figure 7.5 provides the MUST error report for this example, which highlights two issues:

The first error that we report highlights that the initialization of MPI in line 15 only requests `MPI_THREAD_SINGLE`, while the application uses multiple threads. The MPI standard requires that a multi-threaded application specifies a so called *thread*

MUST Output, starting date: Wed Sep 16 12:15:09 2015.

Rank(s)	Type	Message
0(27443)	**Warning**	You requested 4 threads by OMP_NUM_THREADS but requested MPI_THREAD_SINGLE from the …

Details:

Message	From	References
You requested 4 threads by OMP_NUM_THREADS but requested MPI_THREAD_SINGLE from the mpi library. This is ok as long as your application doesn't use any OpenMP between MPI_Init and MPI_Finalize.	Representative location: **MPI_Init_thread** (1st occurrence) called from: #0 main@omptest.c:15	

Rank(s)	Type	Message
0(27453)	**Error**	Error: Thread 3 of 4 (Parallelid: 2) passes a different barrier than other threads of the same team!

Details:

Message	From	References
Error: Thread 3 of 4 (Parallelid: 2) passes a different barrier than other threads of the same team!	Representative location: **main** (1st occurrence) called from: #0 main@omptest.c:15	

Rank(s)	Type	Message
0(27452)	**Error**	Error: Thread 2 of 4 (Parallelid: 2) passes a different barrier than other threads of the same team!
0(27451)	**Error**	Error: Thread 1 of 4 (Parallelid: 2) passes a different barrier than other threads of the same team!
0(27451)	**Error**	Error: Thread 1 of 4 (Parallelid: 2) passes a different barrier than other threads of the same team!
0(27452)	**Error**	Error: Thread 2 of 4 (Parallelid: 2) passes a different barrier than other threads of the same team!
0(27453)	**Error**	Error: Thread 3 of 4 (Parallelid: 2) passes a different barrier than other threads of the same team!

MUST has completed successfully, end date: Wed Sep 16 12:15:11 2015.

Fig. 7.5 MUST error message for the incorrect example application from Listing 7.1

level. Since the example uses `MPI_Init`—instead of `MPI_Init_thread`—the user must assume the lowest thread level that prohibits the use of multiple threads per MPI process. In our example at least `MPI_THREAD_FUNNELED` has to be requested, which means that the process may be multi-threaded, but only the main thread will invoke MPI calls.

The second issue that we report shows the incorrect invocation of the OpenMP barriers. Our checks include the implicit barriers that the `parallel` construct includes, thus, we report multiple instances of the issue.

Our example starts one tool thread per application process and at least one additional tool process, both use memory in the range of several kilobytes. The time overhead for this example error is low since no memory tracing nor any binary instrumentation is necessary. Furthermore, OMPT is designed as a low overhead API that only invokes one additional function call (callback) per OpenMP directive. In addition, MUST stores each event as soon as it observes it and each thread checks whether the other threads of the same team are in a different barrier. We expect that correctness analyses that require memory tracing will have significantly higher overhead.

7.6 Conclusion and Outlook

Increasing single node parallelism of current HPC systems leads to an increased combination of message passing and multi-threading. Especially, the combination of the de-facto standards MPI and OpenMP is common and allows the development of highly scalable applications in many fields of research. However, the use of such hybrid parallel programming paradigms also introduces new classes of usage errors. In order to lower the debugging burden and to increase portability across platforms, we extend the runtime correctness checking tool MUST to include initial support for hybrid applications. Our extensions include thread-safety support and a new communication system to efficiently analyze correctness information from OpenMP threads.

Furthermore, we implement and evaluate first hybrid correctness checks on the basis of an instrumented OpenMP runtime. We highlight that the "OpenMP Tools Application Programming Interface for Performance Analysis" (OMPT) is suitable to check different types of OpenMP usage errors also. Our extensions to the OMPT specification and its prototype implementation in the Intel/LLVM runtime are available as open source.

We target further correctness checks for future extensions of MUST to support a wider range of usage errors from our classification. With the presented work we lay a solid foundation for such a development. Complex correctness checks such as data race detection or checks for the OpenMP `target` directive require information on memory accesses, e.g., provided by binary instrumentation. We plan to integrate such a component into the MUST infrastructure in the future.

Acknowledgments Parts of this work was funded by the German Federal Ministry of Research and Education (BMBF) under Grant Number 01IH13008A (ELP).

References

1. Cramer, T., Dietrich, R., Terboven, C., Müller, M.S., Nagel, W.E.: Performance analysis for target devices with the OpenMP tools interface. In: IEEE International Parallel and Distributed Processing Symposium Workshop (IPDPSW), pp. 215–224. IEEE (2015)
2. Eichenberger, A.E., Mellor-Crummey, J.M., Schulz, M., Wong, M., Copty, N., Dietrich, R., Liu, X., Loh, E., Lorenz, D.: OMPT: an OpenMP tools application programming interface for performance analysis. In: Rendell, A.P., Chapman, B.M., Müller, M.S. (eds.) IWOMP. Lecture Notes in Computer Science, vol. 8122, pp. 171–185. Springer, New York (2013)
3. Eichenberger, A.E., Mellor-Crummey, J., Schulz, M., Copty, N., Cownie, J., Cramer, T., Dietrich, R., Liu, X., Loh, E., Lorenz, D.: OMPT: an OpenMP tools application programming interface for performance analysis. Revised OpenMP Technical Report 2 (2015). https://github.com/OpenMPToolsInterface/OMPT-Technical-Report
4. Hilbrich, T., Schulz, M., de Supinski, B.R., Müller, M.S.: MUST: A Scalable Approach to Runtime Error Detection in MPI Programs. In: Müller, M.S., Resch, M.M., Schulz, A., Nagel, W.E. (eds.) Tools for High Performance Computing: Proceedings of the 2nd International Workshop on Parallel Tools for High Performance Computing. ZIH, Springer Publishing Company, Incorporated (2009)
5. Hilbrich, T., Müller, M.S., de Supinski, B.R., Schulz, M., Nagel, W.E.: GTI: a generic tools infrastructure for event-based tools in parallel systems. In: Proceedings of the IEEE 26th International Parallel and Distributed Processing Symposium, IPDPS'12, pp. 1364–1375. IEEE Computer Society, Washington, DC, USA (2012)
6. Humphrey, W.S.: Bugs or Defects? http://www.uqac.ca/flemieux/PRO102/watts-mar99.pdf (1999), 10 October 2015

7. Liu, X., Mellor-Crummey, J., Fagan, M.: A new approach for performance analysis of OpenMP programs. In: Proceedings of the 27th International ACM Conference on International Conference on Supercomputing, ICS'13, pp. 69–80. ACM, New York, USA (2013)
8. Lorenz, D., Dietrich, R., Tschüter, R., Wolf, F.: A comparison between OPARI2 and the OpenMP tools interface in the context of Score-P. In: DeRose, L., de Supinski, B.R., Olivier, S., Chapman, B., Müller, M. (eds.) Using and Improving OpenMP for Devices, Tasks, and More. Lecture Notes in Computer Science, vol. 8766, pp. 161–172. Springer International Publishing, Cham (2014)
9. MPI Forum: MPI: A Message-Passing Interface Standard, Version 3.0 (2012)
10. Münchhalfen, J.F., Hilbrich, T., Protze, J., Terboven, C., Müller, M.S.: Classification of common errors in OpenMP applications. In: De Rose, L., de Supinski, B.R., Olivier, S.L., Chapman, B.M., Müller, M.S. (eds.) Using and Improving OpenMP for Devices, Tasks, and More. Lecture Notes in Computer Science, vol. 8766, pp. 58–72. Springer International Publishing, Cham (2014)
11. OpenMP Architecture Review Board: OpenMP Application Program Interface, Version 4.0 (2013)
12. Petersen, P., Shah, S.: OpenMP support in the Intel Thread Checker. OpenMP Shared Memory Parallel Programming. Lecture Notes in Computer Science, vol. 2716, pp. 1–12. Springer, Berlin (2003)
13. Schulz, M., de Supinski, B.R.: Pnmpi tools: a whole lot greater than the sum of their parts. In: Proceedings of the ACM/IEEE Conference on Supercomputing, SC'07, pp. 30:1–30:10. ACM, New York, USA (2007)
14. Süß, M., Leopold, C.: Common mistakes in OpenMP and how to avoid them: a collection of best practices. In: Proceedings of the 2005 and 2006 International Conference on OpenMP Shared Memory Parallel Programming, IWOMP'05/IWOMP'06, pp. 312–323. Springer, Berlin (2008)
15. Zeller, A.: Why Programs Fail: A Guide to Systematic Debugging. Morgan Kaufmann Publishers Inc., San Francisco (2005)

Chapter 8
Event Flow Graphs for MPI Performance Monitoring and Analysis

Xavier Aguilar, Karl Fürlinger and Erwin Laure

Abstract Classical performance analysis methodologies use either execution traces with fine-grained data or profiles with aggregates. Event traces provide the finest level of detail on application behavior, however, they are infeasible at extreme scales due to the huge amount of information they contain. In contrast, profiles are much more scalable but they lack the temporal order between events. In this paper, we present the use of event flow graphs for performance characterization of MPI applications. Event flow graphs capture statistics on the events performed by the application while preserving the temporal order of such events. Thus, they stand in between tracing and profiling, and are a good complement for classical performance analysis approaches such as event traces or profiles. Furthermore, event flow graphs can be used for means other than visual exploration of performance data. For example, graphs can be utilized as compressed representations of event traces, or to automatically detect the program structure of iterative applications at runtime without any source code analysis.

8.1 Introduction

Comprehending software behavior is an essential task in the field of High Performance Computing. Parallel computing infrastructures are expensive to maintain and require optimized applications that waste as little computing resources as possible.

Tools for performance analysis of HPC systems and applications have been widely studied for over 20 years. These tools follow mainly two different approaches when collecting and presenting performance data from parallel applications: profiling and

---

X. Aguilar (✉) · E. Laure
High Performance Computing and Visualization Department (HPCViz),
and Swedish E-Science Research Center (SeRC),
KTH Royal Institute of Technology, Lindstedvägen 5, 10044 Stockholm, Sweden
e-mail: xaguilar@pdc.kth.se

K. Fürlinger
Computer Science Department, MNM Team, Ludwig-Maximilians-Universität (LMU)
Munich, Oettingenstr. 67, 80538 Munich, Germany

A. Knüpfer et al. (eds.), *Tools for High Performance Computing 2015*,
DOI 10.1007/978-3-319-39589-0_8

tracing. Profiling consists in generating reports with statistics or aggregates about performance and program trends, usually discarding the temporal order of the events profiled. Although this is a scalable solution for performance monitoring, certain bottlenecks that depend on the event order such as the late-sender problem are difficult to detect with profiles only. On the other hand, tracing generates huge event logs with all application events ordered in time with their corresponding timestamps. While tracers provide a finer level of performance detail, they have limited scalability due to the huge amount of data contained in such traces. For example, tracing one minute of execution for the LU NAS Parallel Benchmark [6] with 128 MPI processes generates one trace file of 2.1 GB.

In [2, 8] the authors presented a novel approach for performance characterization of MPI programs that falls in between profiling and tracing. This approach uses *Event Flow Graphs* in order to achieve the scalability of profiling methods while maintaining the temporal order of application events as in tracing. This paper presents our research up to present time in the use of event flow graphs for MPI performance monitoring and characterization. The remainder of the paper is structured as follows: Sects. 8.2 and 8.3 provide background on event flow graphs and temporal event flow graphs. Section 8.4 presents three different use cases for our event flow graphs: trace compression, automatic detection of application structure, and visual exploration of performance data. Section 8.5 outlines related work. Finally, Sects. 8.6 and 8.7 discuss future work and conclusion, respectively.

8.2 Event Flow Graphs (EFGs)

Our work with event flow graphs builds on top of the Integrated Performance Monitoring (IPM) tool [5, 12], a profiling tool for parallel applications aimed at providing reports on program events such as MPI operations, OpenMP regions or CUDA code. At the moment, we focus our work in MPI events only. Upon program termination, IPM is able to generate for each MPI process a weighted directed graph in which nodes are the different MPI calls performed by the process, and edges the transition between those MPI calls. In other words, graph edges represent the program code blocks executed between two MPI calls. Figure 8.1 presents a simple MPI program and the corresponding event flow graph for one of its processes. As it can be seen in the picture, the event flow graph captures the execution flow of the application. Thus, the graph can be used to reproduce the ordered sequence of events performed by the process without storing any explicit time information such as timestamps.

In addition, the event flow graphs implemented in IPM are also populated with performance statistics and aggregates, for instance, graph nodes can contain aggregates about call duration, and edges can include performance counters about the computation phase represented by such an edge. Therefore, our event flow graphs can capture not only the execution flow of a program but its performance behavior too.

```
void main(int argc, char *argv[ ]) {
    MPI_Init(…);

    for( i = 0; i < 10; i++) {
     MPI_Send(...);
     MPI_Recv(...);
    }
    MPI_Finalize( );
}
```

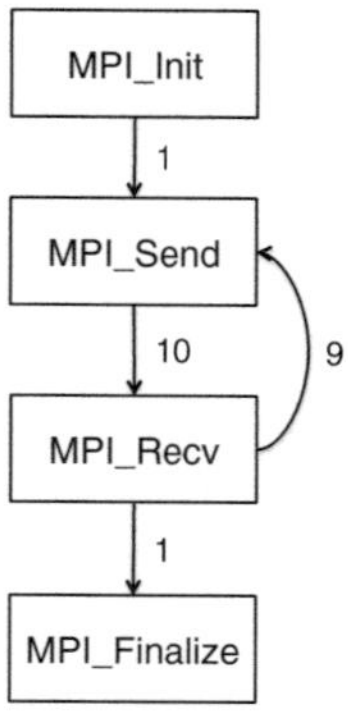

Fig. 8.1 MPI application and its corresponding event flow graph for each MPI process

8.3 Temporal Event Flow Graphs (t-EFGs)

As previously introduced in Sect. 8.2, the full sequence of events from a process can be recovered by traversing its event flow graph from the initial to the final node. Nevertheless, there are cases that cannot be reconstructed using the graph model previously presented. Figure 8.2a shows an application that has a loop containing a conditional branch. If the iteration number is odd, then the process receives data, otherwise, it sends data. In such a case, the event flow graph generated does not contain enough information to recover the order of MPI calls across iterations as depicted in Fig. 8.2b. Therefore, we developed *Temporal Event Flow Graphs*, a modified version of our event flow graphs that guarantees trace recovery [2].

(a)

```
int main( ) {
    MPI_Init(…);
    for(i = 1; i <= 10; i++) {
     MPI_Bcast(...);
     if(i%2) /* odd */
       MPI_Recv(...);
     else  /* even */
       MPI_Send(...);
    }
    MPI_Finalize( );
}
```

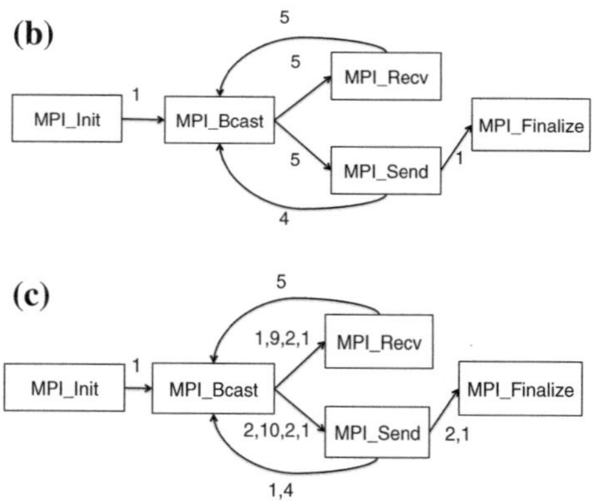

Fig. 8.2 MPI code and its corresponding event flow graph and temporal event flow graph. **a** MPI code with a conditional branch in-side a loop. **b** Event flow graph where trace recovery cannot be guaranteed. **c** Temporal event flow graph with sequence information on edges

Temporal event flow graphs keep track of the outgoing edge order in nodes with more than one outgoing edge. This information is stored in a compact way in the edge labels. Edge labels can have two different formats: $< \mathbb{N}, \mathbb{N} >$ or $< \mathbb{N}, \mathbb{N}, \mathbb{N}, \mathbb{N} >$. The first case represents the simplest case, that is, an edge with a single sequence number. For example $< 2, 1 >$, meaning that this edge is taken once the second time we traverse the node. In contrast, the second case defines a partition of the iteration space performed by the application in the form of *<first sequence number, last sequence number, number stride, edge count>*. Figure 8.2c depicts the temporal event flow graph generated from the code in Fig. 8.2a. Now with this new graph model, trace recovery is always guaranteed because the graph keeps for every edge the sequence order in which it has to be traversed.

8.4 Three Different Use Case Scenarios for Event Flow Graphs

In this section we present three uses cases for employing event flow graphs. First we describe their usage for compressing traces, second we show how to exploit them to detect the loop nesting structure of applications and finally, we demonstrate their utility for visual performance analysis. In order to do so, we used several mini-applications that represent typical HPC scientific codes. We utilized different codes from several benchmark suites to have a wide representation of HPC codes such as stencils, conjugate gradients, multigrid solvers, finite element codes, etc.

8.4.1 Event Flow Graphs and Trace Compression

Temporal event flow graphs record the order of application events with very low overhead in terms of time and space. By using graphs, the amount of performance data stored is reduced as program events and most of the structural data (like e.g. sender/receiver information) are encoded in the graph, and not repeated several times, which is typically done in traces. Therefore, our temporal event flow graphs can be considered as a compressed representations of event traces [2].

In our first experiment, we measured the overhead introduced into the test applications when collecting and storing event flow graphs into disk. For all of them, the overhead introduced when running strong and weak scaling tests up to 800 cores was less than 2 % of total running time. In other words, only 2 % of the total program time was spent in intercepting MPI calls, building the graphs and writing them to disc. Furthermore, graph creation is performed locally per process, thus, it is independent to the number of cores used in the application.

Our second set of experiments consisted in measuring the compression ratios between temporal event flow graph and event traces. These ratios were computed

Table 8.1 Compression results achieved by using event flow graphs on several mini-application benchmark codes

Benchmark	AMG	GTC	MILC	SNAP	MiniDFT	MiniFE	MiniGhost
#Ranks	96	64	96	96	40	144	96
Comp. Factor	1.76	46.60	39.03	119.23	4.33	19.93	4.85

comparing standard event traces generated by IPM with their respective event flow graphs. Traces and graphs contained exactly the same amount of information. For each MPI call this included the call name, bytes sent or received, communication partner, and call site. Timestamps were not included in the experiments as time-stamp reconstruction remains as future work. It is important to remark that each trace could be exactly reconstructed from their corresponding event flow graph.

Table 8.1 shows the compression ratios obtained for various benchmarks. These ratios depend on the application and range from 1.79x up to 119x. In terms of file size, a 119x compression ratio implies that the space required to store the temporal event flow graphs for SNAP was only 10 MB, compared to 1.1 GB required for the complete set of regular trace files. This gain comes from the fact that graphs can encode structural data such as loops of events that do not need to be repeated for every event as typically done with tracers. As can be seen in the table, the compression factor depends on the graph generated from the application. Graphs with a small number of nodes with high frequency (how many times a node was executed by the program) lead to higher compression ratios.

8.4.2 Event Flow Graphs and Application Structure Detection

Temporal event flow graphs can also be used for application structure detection. Our approach for automatic analysis of flow graphs sheds light on the inherent loop nesting structure of MPI parallel applications [3]. Knowledge of this loop nesting structure is very useful for postmortem and for on-line performance analysis purposes. On one hand, it can be utilized to automatically generate reports that show statistics about loops without any source code instrumentation. On the other hand, knowing the structure of a program while it runs can be beneficial to collect data in a more efficient way. For instance, data can be aggregated at a loop level instead of keeping every event, or fine-grain data can be only collected for a few loop iterations.

It is commonly accepted that the vast majority of HPC scientific parallel codes are iterative and spend most of their time in loops. These applications are usually composed of a large outer loop, which drives the simulation time-steps, and which contains several inner loops with different nesting levels. Table 8.2 shows some statistics about loops for several scientific codes. The table contains the total time for each

Table 8.2 Loop statistics for several benchmark codes

Benchmark	#Ranks	Total time (s)	Outermost loop(s) count	Time in all outermost loop(s) (%)	Time in dominant loop (%)
MiniGhost	96	282.17	1	98.8	98.8
MiniFE	144	133.50	13	78.1	77.7
BT	144	370.59	7	99.4	99.0
LU	128	347.53	3	99.2	98.9

application, the number of outermost loops (loops not nested within other loops), and the percentage of total running time (inclusive) spent in those outermost loops. Dominant loop refers to the most time consuming outermost loop. As it can be seen in the table, the big outer loop hypothesis largely holds for these examples.

Most MPI parallel programs contain MPI operations in some of their loops in order to share data among processes across simulation iterations. In such cases, the generated temporal event flow graphs will contain cycles. Thus, by detecting the graph cycles, we are detecting the actual loops that drive the simulation in the application. Loops without MPI calls will not generate cycles in the t-EFG, however, their behavior gets captured in the edges of the graph since those could be considered as simple computational blocks between two MPI calls.

Algorithms for cycle detection in graphs have been widely studied in the field of compilers [11, 20, 22]. Our analysis framework implements an efficient DFS-based algorithm presented in [23]. This algorithm runs in almost linear time and does not require any complicated data structures unlike other cycle detection algorithms. Therefore, it is much easier to implement.

After algorithm finalization, all graph cycles are detected and nodes are labeled accordingly to the following criteria: loop header nodes (entry node in a graph cycle) are identified and marked, and all nodes within a loop are labeled with their corresponding loop header node. Thereby, our analysis framework can build a hierarchical loop tree as shown in Fig. 8.3b. The tree root represents the process, every internal node a loop, and leaves are the program events. As can be seen in the figure, the tree depicts that the graph has two nested loops with headers *MPI_2* and *MPI_3*. This tree models naturally the hierarchical structure of loops in the program, and allows our monitoring infrastructure, among other things, to know the number of cycles in the graph, if two cycles are nested, if an event belongs to a certain cycle, etc.

8.4.2.1 Loop Detection at Runtime

MPI scientific applications are usually iterative programs that exhibit a stable performance behavior across iterations during execution time, that is, the main simulation loop (and its sub loops) have very similar performance during the whole application

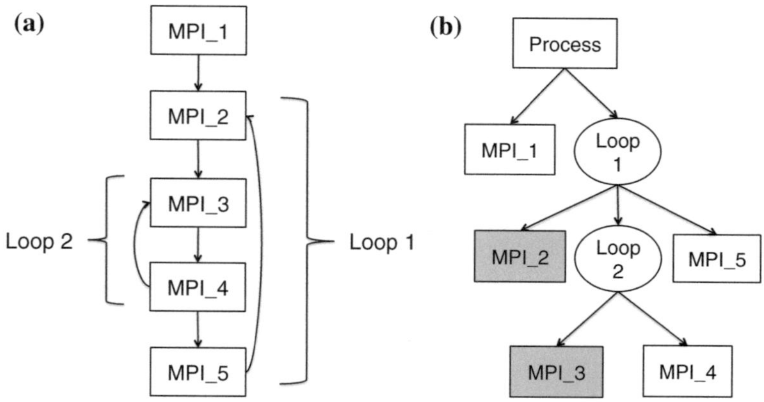

Fig. 8.3 Graph with two nested loops and its corresponding nesting loop tree. **a** Graph with two nested loops. **b** Nesting loop tree

lifetime. Thus, keeping information on only a few iterations should be sufficient to characterize the overall performance behavior of the application.

Our graph cycle analysis can be performed at runtime, thereby, being able to identify the main simulation loop to collect fine-grained information of only a few chosen iterations.

In order to minimize the amount of overhead introduced, the graph cycle detection is triggered only once the application has reached a stable state. An application is considered stable when it enters an iterative phase where its performance behavior shows only minor fluctuations. Most scientific codes reach this state when they start executing the main simulation loop. In our case, this situation can be detected in the number of nodes in the temporal event flow graph. Once the application enters its iterative phase, the number of nodes in the graph does not change because the application executes the same MPI calls again and again.

Table 8.3 compares the sizes of whole-run traces against traces with only 10 iterations automatically collected with our tool. As can be observed in the table, the trace size can be reduced by several orders of magnitude by keeping information on only a few iterations. Nevertheless, keeping information for a few iterations always comes with an inevitable data loss. Therefore, we also checked if those few selected

Table 8.3 Size comparison between whole execution traces and traces with 10 selected iterations for several benchmark codes

Metric	MiniGhost	MiniFE	MiniMD	GTC	BT	LU
Full trace	26 MB	77 MB	555 MB	48 MB	717 MB	7.7 GB
10-iterations	4.4 MB	4.1 MB	788 KB	1.3 MB	29 MB	267 MB
% reduced	83 %	94.7 %	99.8 %	97.3 %	96 %	96.53 %

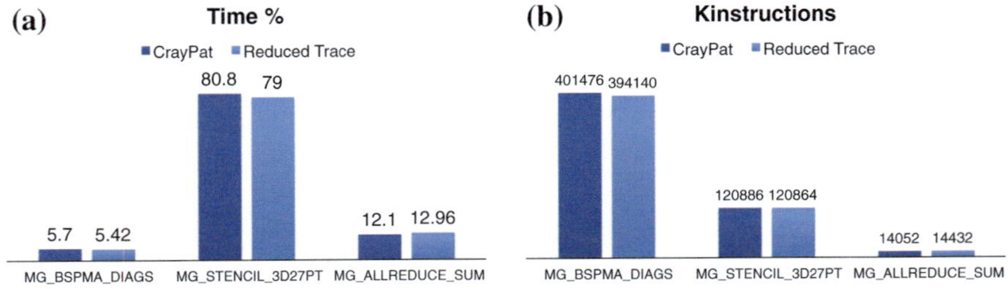

Fig. 8.4 Statistics from CrayPat for a whole run of MiniGhost compared to statics generated from a 10-iterations trace. **a** Percentage of time spent in the three most time consuming functions. **b** Kilo instructions executed in the three most time consuming functions

iterations still captured the overall behavior of the application for its total running time. To that end, we used CrayPat to identify the most time consuming functions in MiniGhost (an MPI stencil mini-application), and afterwards, we compared profiles generated by CrayPat for those three functions with statistics computed from our 10 iteration trace. Figure 8.4 shows that the percentage of time and the number of instructions executed in the three most time consuming functions of MiniGhost are very similar when we compare the whole run with the 10 iterations trace. Thus, we can conclude that the trace with only 10 iterations keeps the overall performance behavior for the whole run of MiniGhost in 83 % less space than the original trace.

Our current approach collects data for a few consecutive iterations only once during the whole lifetime of the application, and thus, it leaves room for some improvements. It is planned in our future work to take other metrics such as performance hardware counters into account to trigger the data collection. Thereby, being able to collect various sets of iterations during the same run if the application degrades over time or if it has irregular behavior.

8.4.2.2 Tree Clustering to Detect Process Similarity

Generating loop nesting trees is not only useful to unveil loop nesting structures at a process level but also to compare between processes. In our case, two processes have similar behavior if their loop nesting trees are isomorphic, that is, they have the same height and the same number of nodes per level. If two processes have isomorphic trees with the same events in their leaves, we can guarantee that such processes have the same number of loops, with the same structure, and that they have executed the same calls with the same call paths.

Clustering processes with the same behavior into groups is beneficial to increase tool scalability by reducing the amount of redundant data collected. Once groups of processes are known, performance information can be collected from one representative of each group instead of collecting data from all of them.

In order to explore this tree clustering approach, we implemented an algorithm that groups hierarchical loop trees from an application into clusters. Thereby, detecting groups of processes with similar behavior within the application. For each pair of processes, the algorithm checks if their trees are isomorphic. In other words, it analyses if two trees have the same shape, and if there is a mapping of MPI call name and call site between leaves of such two trees. Transfer size or communication partner rank are not included in the comparison since different processes execute the same MPI operations with different communication partners, and with different transfer sizes. Once the trees are compared, they are grouped in the same cluster if they are isomorphic.

We tested our clustering approach with the NAS Parallel Benchmarks (NPB) and several mini-applications from the Mantevo Project. Due to space limitations, we restrict the analysis presented here to MiniGhost from the Mantevo Project.

We run MiniGhost with 96 processes calculating a 3D 7-point stencil. In this case, our clustering algorithm obtains 8 different clusters: four of them including 8 processes each, and the remaining four, 16 processes respectively. This cluster decomposition can be explained by the logical processor grid used in the experiments. Figure 8.5a shows this grid of $4 \times 4 \times 6$ processors colored by cluster. Clusters 1, 2, 3, 4 are present in the front and back planes of the grid, whereas clusters 5, 6, 7, 8 include only processors from the inner planes. Figure 8.5b depicts the clusters mapped into the front and back planes of the grid. There are 4 different clusters with 4 processors each: processors in the corners belong to one cluster as they communicate in the same manner with 3 neighbors (one in the north-south direction, one in the east-west, and one in the front-back); accordingly, the 4 central processors belong to another cluster as they communicate with 5 neighbors (north, south, east, west, and one in the front-back direction); finally, processors in the borders of the plane are in another two different clusters. In this case, although processes in cluster 3 and 4 communicate with the same number of neighbors, they are grouped into two separate clusters due to their graph structure. Their number of MPI calls is the same,

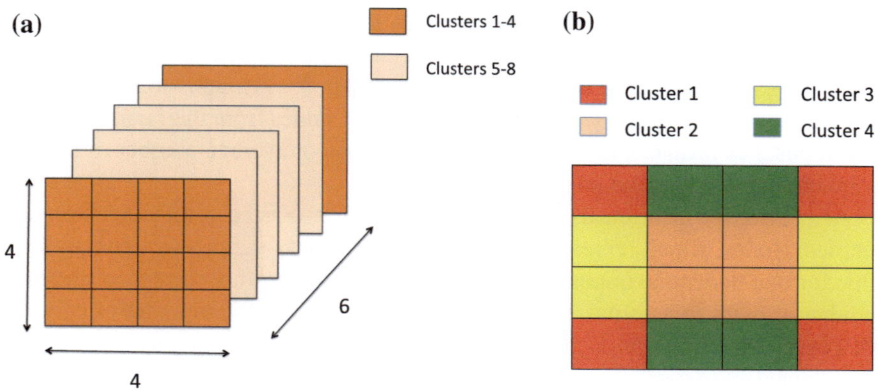

Fig. 8.5 Cluster decomposition for the MiniGhost application. **a** Clusters mapped on the logical processor grid used in MiniGhost. **b** Mapped clusters in the front and back planes of the processor grid

but their sequence diverges because the number of neighbors is different regarding the direction. Processors in cluster 3 have two neighbors in the north-south direction whereas processors in cluster 4 have two in the east-west direction. Thus, the order of MPI calls is different, and as a result, their event flow graphs are different too. For the processors in the 4 inner planes of the cube, we have again the same mapping but with clusters 5, 6, 7, and 8. These inner processes do not belong to clusters 1, 2, 3, or 4 because they have one additional neighbor in the front-back direction.

8.4.3 Event Flow Graphs and Visual Performance Analysis

Temporal event flow graphs can also be used in the task of visual performance analysis of MPI applications. As previously discussed, our graphs include performance statistics about the application behavior together with visual information about the inherent order of program events. Thus, they can be used as an intermediate step between profiles and traces. For instance, a performance analyst could start looking at several profiles to get an overview on the performance of an application. Then, if he wants more detail, he could get down into event flow graphs to look at those statistics with a finer level of detail and with the temporal factor added, that is, how different events and metrics are placed in time. Finally, the analyst could even have a greater level of detail by using traces in order to see what happened exactly at every single moment in time.

Nevertheless, event flow graphs from real applications can be huge and complex, therefore, we implemented in [4] two different techniques to help the analyst in the task of performance exploration using graphs: automatic cycle compression and graph coloring.

Automatic cycle detection consists in automatically detecting cycles in graphs and compressing each one of them into a different single node. As a result, the number of nodes in the graph is reduced and the graph readability increases. The user can click one of this loop nodes and open a hyper-linked subgraph containing only the nodes an edges corresponding to that loop. Furthermore, loop nodes in graphs are also useful to summarize statistics as they can show accumulated values from their corresponding subgraphs. For instance, total time spent in that loop or average number of floating point operations per iteration.

As previously mentioned, graph elements have associated diverse performance metrics. Nodes, which represent MPI calls performed by the process, contain total time for that node, maximum and minimum time, number of occurrences, and total number of bytes transferred. For the edges, which are computation phases between MPI calls, different timings and hardware counters using the PAPI interface [16] can be collected and saved. These performance metrics can be used to color the graph in order to highlight the most interesting parts of it. For instance, coloring graph nodes with a gradient representing the amount of time spent in them, or coloring the edges using hardware counters information to emphasize computational phases with bad single-core performance. Metrics that can be currently used for event flow graph coloring are time, bytes and number of occurrences for nodes; and time, instructions per cycle (IPC), million of instructions per second (MIPS), million of floating point operations per second (MFLOPS) and cache misses for edges.

8.5 Related Work

There is a wide variety of performance tools to monitor and analyze parallel applications. However, and to our knowledge, no previous work in the use of flow graphs for performance monitoring and characterization exists. Paraver [19], Vampir [17] and Jumpshot [24] are visualization frameworks for the analysis of event traces. They provide exploration of performance data with the greatest level of detail possible. Other tool frameworks such as HPCtoolkit [1], Scalasca [9], and TAU [21] support profiling and tracing at the same time, although their strength remains in the exploration of performance profiles. In this context, our event flow graphs stand between these two families of tools because our graphs are not traces nor pure profiles. Therefore, our work fits perfectly as a complement for all these existing performance analysis frameworks.

Detection of application structure is the topic of several related works. The works of Gonzalez et al. [10], and Casas et al. [7] use burst clustering and spectral analysis techniques respectively to detect program structure in event traces. Although they are very good approaches they either require whole event traces or complex parallel software infrastructures to be performed in an on-line scenario [15]. In contrast, our approach can be performed on-line with very low overhead and without the need of having full event traces in advance.

Other works have explored as well the compression of event traces for performance analysis. Scalatrace [18] is a tracing framework that provides on-the-fly lossless trace compression of MPI communication traces. Krishnamoorthy et al. [14] use SEQUITUR for trace compression by creating context free grammars from the sequence of MPI calls. Knüpfer et al. use Complete Call Graphs (CCGs) for post-mortem trace compression [13]. Their approach relies in building call trees and replacing similar repeated sub-trees with references to a single instance. Our approach differs from the others in the fact that our event flow graphs have a more general purpose, and can be used for many other things than trace compression.

8.6 Future Work

Our current implementation of temporal event flow graphs focuses in MPI applications, however, upcoming hybrid architectures and fat computing nodes empower the use of hybrid approaches in parallel programming. Thus, we plan to extend our flow graphs adding the possibility to model and characterize hybrid codes that use MPI+OpenMP, or MPI+PGAS for example.

Another aspect we will explore in the future is the use of event flow graphs to reconstruct event traces containing continuous data, i.e. time-stamped information or hardware performance counters.

Finally, we want to continue investigating visualization techniques to use event flow graphs in the task of visual performance analysis. For example, automatic methods to highlight differences between graphs, or semantic view aggregation techniques to visualize different graphs at the same time.

8.7 Conclusion

The use of event flow graphs for performance monitoring and characterization opens up many new possibilities as presented in this paper. First, temporal event flow graphs are useful to store performance data in a compressed format. Second, event flow graphs can be used to automatically detect the loop nesting structure of applications without the need of any source code analysis. Knowing the structure of an application is useful, for instance, to automatically compute post-mortem statistics about loops, or to perform smart data collection techniques such as tracing automatically a subset of iterations of the main program loop. Finally, event flow graphs can be also used in the task of visual performance analysis. Automatic graph cycle detection together with graph coloring provides an intuitive mechanism to explore application performance data, visualizing the temporal relations between the different events executed by the program.

References

1. Adhianto L, Banerjee S, Fagan M, Krentel M, Marin G, Mellor-Crummey J, Tallent NR (2010) Hpctoolkit: Tools for performance analysis of optimized parallel programs. Concurr Comput Pract Exp. 22(6):685–701
2. Aguilar X, Fürlinger K, Laure E (2014) MPI trace compression using event flow graphs. In: Proceedings of the Euro-Par 2014 parallel processing, Springer, Heidelberg, pp 1–12
3. Aguilar X, Fürlinger K, Laure E (2015) Automatic on-line detection of MPI application structure with event flow graphs. In: Proceedings of the Euro-Par 2015: Parallel processing, Springer, Heidelberg, pp 70–81
4. Aguilar X, Fürlinger K, Laure E (2015) Visual MPI performance analysis using event flow graphs. Procedia Comput. Sci. 51(0), 1353–1362, In: Proceedings of the international conference on computational science ICCS 2015, Computational science at the gates of nature
5. Aguilar X, Laure E, Furlinger K (2013) Online performance data introspection with ipm. In: Proceedings of the high performance computing and communications & IEEE international conference on embedded and ubiquitous computing (HPCC_EUC), IEEE 10th international conference, IEEE, pp 728–734
6. Bailey DH, Barszcz E, Barton JT, Browning DS, Carter RL, Dagum L, Fatoohi RA, Frederickson, PO, Lasinski TA, Schreiber RS et al (1991) The nas parallel benchmarks. Int J High Perform Comput Appl 5(3):63–73

7. Casas M, Badia RM, Labarta J (2010) Automatic phase detection and structure extraction of MPI applications. Int J High Perform Comput Appl 24(3):335–360
8. Fürlinger K, Skinner D (2010) Capturing and visualizing event flow graphs of MPI applications. In: Proceedings of the Euro-Par 2009–parallel processing workshops. Springer, Heidelberg, pp 218–227
9. Geimer M, Wolf F, Wylie BJ, Ábrahám E, Becker D, Mohr B (2010) The scalasca performance toolset architecture. Concurr Comput Pract. Exp 22(6):702–719
10. Gonzalez J, Gimenez J, Labarta J (2009) Automatic detection of parallel applications computation phases. In: Proceedings of the parallel distributed processing on IEEE international symposium IDPDS 2009, pp 1–11
11. Havlak P (1997) Nesting of reducible and irreducible loops. ACM Trans Progr Lang Syst (TOPLAS) 19(4):557–567
12. IPM WWW site: http://www.ipm2.org
13. Knupfer A, Nagel WE (2005) Construction and compression of complete call graphs for post-mortem program trace analysis. In: Proceedings of the international conference on parallel Processing ICPP 2005, IEEE, pp 165–172
14. Krishnamoorthy S, Agarwal K (2010) Scalable communication trace compression. In: Proceedings of the 2010 10th IEEE/ACM international conference on cluster, cloud and grid computing. IEEE Computer Society, pp 408–417
15. Llort G, Gonzalez J, Servat H, Gimenez J, Labarta J (2010) On-line detection of large-scale parallel application's structure. In: Proceedings of the IEEE international symposium on parallel distributed processing (IPDPS), pp 1–10
16. Mucci PJ, Browne S, Deane C, Ho G (1999) Papi: a portable interface to hardware performance counters. In: Proceedings of the department of defense HPCMP users group conference, pp 7–10
17. Nagel WE, Arnold A, Weber M, Hoppe HC, Solchenbach K (1996) Vampir: visualization and analysis of mpi resources
18. Noeth M, Ratn P, Mueller F, Schulz M, de Supinski BR (2009) Scalatrace: scalable compression and replay of communication traces for high-performance computing. J Parallel Distrib Comput 69(8):696–710
19. Pillet V, Labarta J, Cortes T, Girona S (1995) Paraver: a tool to visualize and analyze parallel code. In: Proceedings of WoTUG-18: transputer and occam developments. vol 44, pp 17–31
20. Ramalingam G (1999) Identifying loops in almost linear time. ACM Trans Program Lang Syst (TOPLAS) 21(2):175–188
21. Shende SS, Malony AD (2006) The tau parallel performance system. Int J High Perform Comput Appl 20(2):287–311
22. Sreedhar VC, Gao GR, Lee YF (1996) Identifying loops using DJ graphs. ACM Trans Program Lang Syst (TOPLAS) 18(6):649–658
23. Wei T, Mao J, Zou W, Chen Y (2007) A new algorithm for identifying loops in decompilation. In: Nielson H, File G (eds) Static analysis. Lecture notes in computer science, vol 4634. Springer, Heidelberg, pp 170–183
24. Zaki O, Lusk E, Gropp W, Swider D (1999) Toward scalable performance visualization with jumpshot. Int J High Perform Comput Appl 13(3):277–288

Chapter 9
Aura: A Flexible Dataflow Engine for Scalable Data Processing

Tobias Herb, Lauritz Thamsen, Thomas Renner and Odej Kao

Abstract This paper describes Aura, a parallel dataflow engine for analysis of large-scale datasets on commodity clusters. Aura allows to compose program plans from relational operators and second-order functions, provides automatic program parallelization and optimization, and is a scalable and efficient runtime. Furthermore, Aura provides dedicated support for control flow, allowing advanced analysis programs to be executed as a single dataflow job. This way, it is not necessary to express, for example, data preprocessing, iterative algorithms, or even logic that depends on the outcome of a preceding dataflow as multiple separate jobs. The entire dataflow program is instead handled as one job by the engine, allowing to keep intermediate results in-memory and to consider the entire program during plan optimization to, for example, re-use partitions.

9.1 Introduction

More and larger datasets become available as, for example, sensing device data, user-generated content, and software logs. Gaining insights into this data is becoming increasingly important for many applications. For analyzing this data quickly and efficiently, different scalable data analytic frameworks have been developed. Prominent examples include MapReduce [5], Spark [15, 16], and Flink [1]. In general, these systems share similar objectives, namely hiding the difficulty of parallel programming, providing fault tolerance, and allowing programs to be run efficiently on different cluster infrastructures. Developers write sequential programs and the processing framework takes care of distributing the program among the available compute nodes and executing each instance of the program on appropriate data partitions.

T. Herb (✉) · L. Thamsen · T. Renner · O. Kao
Technische Universitt Berlin, Berlin, Germany
e-mail: tobias.herb@tu-berlin.de

A. Knüpfer et al. (eds.), *Tools for High Performance Computing 2015*,
DOI 10.1007/978-3-319-39589-0_9

In most frameworks, the execution plan is static. Even if some frameworks support iterative programs, users cannot define arbitrary termination criteria. Furthermore, to the best of our knowledge there is no dedicated support for executing one of multiple branches of dataflows based on a condition. At the same time, analytical programs implementing, for example, graph or machine learning algorithms often require control flow. One approach to implement these structures in dataflow systems is to do it on client-side [15, 16]. In this scenario, the client-side driver program creates and submits multiple jobs. However, even if the necessary dataflows jobs are generated from higher-level programming abstractions, executing a program as multiple jobs is often less efficient. It entails full data materialization between jobs, which often means writing these intermediate results to disk. Furthermore, the execution engines only knows about single jobs, not the entire dataflow program, limiting the scope of automatic plan optimization to parts of the programs.

With this motivation, we present Aura, a flexible dataflow engine for scalable data processing. Aura contains a novel mechanism for control flow that enables adjustments of the execution plans at runtime. This allows developers to express more complex analytic programs as a single job, allowing more optimized program execution. Aura realizes control-flow by the interaction between the centralized stepwise evaluation of control-flow logic and the thereby controlled distributed execution of the parallel dataflow.

Outline: The remainder of the paper is organized as follows. Section 9.2 describes related work. The system architecture and its components is presented in Sect. 9.3. Section 9.4 discusses AuraFLs program representation and translation. Section 9.5 presents its distributed execution model. Section 9.6 concludes this paper.

9.2 Related Work

This section describes the evolution of distributed dataflow engines and their programming abstractions. As the execution engines support more general job graphs and features like automatic plan optimization, the programming abstractions become more high-level and declarative.

9.2.1 Distributed Execution Engines for Parallel Dataflows

Aura builds upon the following engines and adds dedicated support for control flow.

MapReduce: The MapReduce [5] paradigm and its implementations allow to execute the two user-defined functions *Map* and *reduce* data-parallely on hundreds of commodity nodes. The execution is fault-tolerant as the results of each execution step are written to a fault-tolerant distributed file system and failing tasks are re-started.

Directed Acyclic Job Graphs: Systems like Dryad [7] and Nephele [13] build upon the data-parallel execution model, but allow more general job graphs than the two alternating functions with in-between data shuffling. As an example, tasks in Nephele can be connected in a general directed acyclic graph in which parallel tasks instances are connected either point-to-point or all-to-all. Furthermore, connections can be disk-based or in-memory.

Relational Operators and Plan Optimization: While tasks in Nephele and Dryad execute arbitrary user code, Scope [4] and Stratosphere/Apache Flink [1, 3] provide a set of pre-defined operators, which include common database operators like joins, but also the second-order functions Map and Reduce. Both Scope and Stratosphere automatically compile and optimize logical query plans to physical plans from SQL-like programming abstractions.

Iterative Dataflows: Some systems, including Flink and Naiad [8, 9] also provide dedicated support for iterative programs, allowing directed cyclic job graphs. Both Flink and Naiad can also process datasets incrementally [6], allowing to exploit the sparse computational dependencies of many iterative algorithms.

Resilient Distributed Datasets: Spark [16] also allows scalable data analysis on commodity clusters, but is based on Resilient Distributed Datasets (RDD) [15]. RDDs are in-memory datasets that are distributed across nodes and on which operations are executed in parallel. The fault-tolerance in Spark does not rely on disk-based replication but on linage, so parts of RDDs can be recomputed when nodes fail.

9.2.2 Programming Abstractions for Parallel Dataflows

The programming abstractions for parallel dataflows are becoming more declarative and more high-level. At the same time, there is a development towards programming languages instead of mere library-based APIs. As such languages include control flow, they provide a direct motivation for supporting control flow on the level of the dataflow engine.

Declarative Programming Abstractions: Pig [10] and Hive [12] are SQL-like programming abstractions for MapReduce. Programs are arguably higher-level and more declarative. In fact, a single Pig or Hive programs usually results in multiple MapReduce jobs. Job compilation, thus, decouples the programming abstraction from actual dataflow jobs. DryadLINQ [14] and the Scope scripting language [4] are similar in that programmers are not limited to the two second-order functions Map and Reduce, but can also use typical relational operations.

Deep Language Embedding: While many programming abstractions for parallel dataflows are library-based, which are often tightly coupled to the underlying execution engine, Emma [2] uses *deep language embedding* to provide much more implicit parallelism. Emma compiles execution plans directly from code—rather than with an API—to enable a more holistic analysis of dataflow programs. Thus enabling additional optimizations, as well as a more complex control flow.

9.3 System Architecture

Figure 9.1 shows an overview of AuraFLs system architecture. As many scalable data processing frameworks [5, 13, 16], the architecture follows the master/worker pattern for parallelizing the execution over many nodes, as visible in (a) of the system overview. In general, Aura consists of three logical core entities:

- the master, called *Workload Manager* (WMs),
- one or more worker instances, called *Task Managers* (TMs), and
- one or more client instances

The following describes the inner structure and responsibilities of these components as well as the interactions between them in more detail.

9.3.1 Workload Manager

The central *Workload Manager* component manages and coordinates each running system instance of Aura. Its inner parts are shown in (c) of Fig. 9.1. A fundamental service of the WM is the cluster management, which entails acquiring, coordinating, and monitoring available cluster resources. For instance, newly available or failed nodes are signaled to all other components by the central cluster manager.

Submitted dataflow programs are received by the WM. Afterwards, the WM internally creates a *Topology Controller* (TC) that takes care of the lifecycle management for the program. The lifecycle includes the parallelization, scheduling, deployment, monitoring and control-flow evaluation of the distributed dataflow program. Intermediate state and aggregated computation results of the executing dataflow are continuously propagated back from the TMs to the corresponding TC. Thus, the TC is able to decide dynamically over the next execution steps of the program, depending on the gathered information. This novel feedback mechanism enables dynamic dataflow execution and allows to express more complex analytical programs than possible with static dataflows.

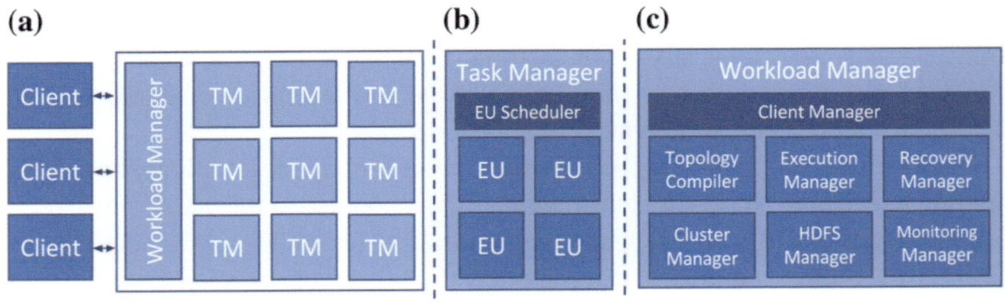

Fig. 9.1 Aura's system architecture: **a** overall master/worker architecture, **b** Task Manager component, and **c** Workload Manager component

Another important function of the WM is the coordinated access to an underlying distributed file system such as the Hadoop Distributed File System (HDFS) [11]. The WM keeps track of all file resource locations and schedules file access to co-located worker nodes to minimize network access. In case of failures, e.g. signaled by the cluster management, the TC triggers the necessary fault-tolerance logic to suspend execution, recover from failure, and resume distributed execution.

9.3.2 Task Manager

Computational tasks of a dataflow program are delegated and scheduled to *Task Managers* (TM). A TM consists essentially of a set of execution units, which perform tasks in parallel, as shown in (b) of Fig. 9.1. Communication of dependent tasks is realized either via in-memory channels for tasks scheduled to different execution units of the same TM, or via network channels for tasks scheduled to execution units of different TMs. A Task Manager has as many *Execution Units* (TM-EU) as cores are available to ensure that each computational task is assigned to a core.

9.3.3 Client

The client is responsible for submitting dataflow programs to a running instance of Aura. It can obtain status updates and possibly failure messages (stack trace) from the WM during program execution. The engine supports parallel programs, submitted from different clients.

9.4 Program Representation and Translation

This section treats the operator plan (i.e. logical plan) and physical plan (i.e. executable program representation) from the perspective of the execution engine. Further, it covers the translation process that generates the physical plan from the operator plan.

9.4.1 Operator Plan

An operator represents a data-parallel function that transforms an individual data record or a collection of data records and returns the result. Typically, data records are ingested as arguments. Our system supports unary operators like map, filter, and aggregate functions as well as binary operators like a join or cartesian product.

These operators are assembled to tree structures expressing a concrete analytical query, the so-called operator plan. The concrete processing takes place by streaming data records from bottom operators—i.e. the leafs in a concrete operator plan—along the edges up to the root operator, which receives the resulting data records of the processed query. The generation of a physical plan is realized by a query language compiler.

Additionally to the set of data-parallel operators we introduce two control-flow operators: (1) *select operator* and (2) *loop operator*.

The select operator acts as an if-statement in common programming languages and allows flexible dataflow execution depending on intermediate state of the current execution. It is parameterized with a predecessor plan, two successor plans, and a condition plan. The select operator first executes the predecessor plan and then the condition plan, which must produce a boolean result. Dependending on this result, the first (true) successor—or the second (false) successor plan is executed.

The loop operator realizes a loop-construct with termination condition like in an imperative programming language. The loop operator is parameterized with a loop plan and an termination plan. First, the loop plan is executed and then the termination plan is executed, which also has to yield a boolean result. Depending on the result the loop plan is re-executed (true) or not (false).

9.4.2 Physical Plan

The physical plan consists of multiple directed acyclic graphs (DAG), called partial physical plan, wired together logically by control-flow operators. The control-flow operators are parameterized with the corresponding partial physical plans and are composed together to a tree structure, comparable to an abstract syntax tree in compilers. Vertices of partial physical plans are distributable program units, containing one or more data-parallel operators. These vertices are deployed to execution units of the TMs. Edges between dependent vertices in the DAG are deployed as physical communication channels (TCP or in-memory). Edges between control-flow operators and vertices exchange only flow-logic information and no data records. An important feature of the channels is the so-called *intermediate cache*. This cache allows to materialize the data stream in memory and/or on disk either on the sender or receiver side. The intermediate cache is fundamental to enable control-flow, because the materialized data must first be evaluated by the condition plan of the control-flow operator and then be re-consumed by the selected successor plan.

9.4.3 Plan Translation and Deployment

The plan translation is the process where the physical plan is generated from the operator plan. In the first step operators are embedded in vertices. Embedding one operator in one execution vertex would lead, in the case of large operator plans, to a very inefficient execution because data records would have to be passed through many network channels. Under certain conditions several connected operators can be compactified to a macro operator which is then embedded in an single vertex. The necessary conditions for this compactification are:

- same partitioning keys,
- same partitioning strategy, and
- same degree of parallelism (DOP).

After the compactification (see Fig. 9.3) step the data-parallel instances of the vertices are generated, according to their DOPs. In the plan scheduling and deployment phase data-parallel vertex instances are assigned and shipped to the corresponding execution units of the TMs (see Fig. 9.2).

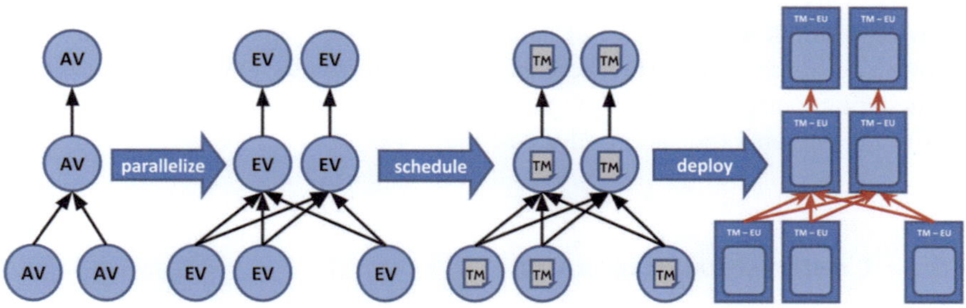

Fig. 9.2 Parallelizing, scheduling, and deployment of a physical plan

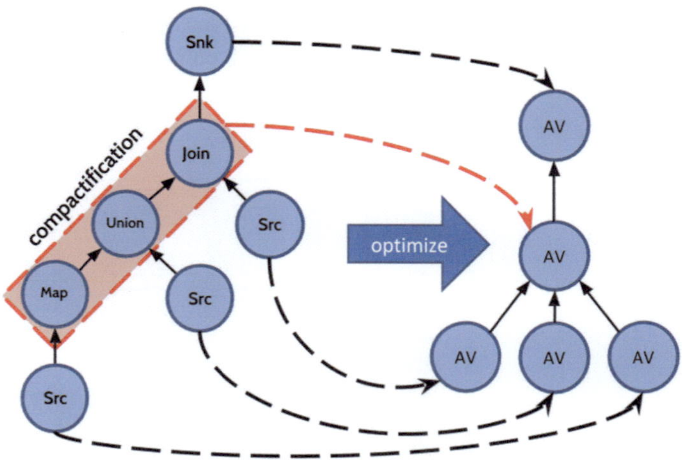

Fig. 9.3 Compactification of operators in the plan translation phase

9.5 Distributed Execution

This section covers the push/pull-based data streams in our distributed execution model, which is shown in Fig. 9.4, and describes the execution of control-flow operators.

9.5.1 *Push/Pull-Based Data Streams*

Data records are either java objects or collections, e.g. sets or lists. For the external data transport, i.e. data transport between vertices, the records are serialized into transport buffers and pushed form sender to the receiver vertices. If the receiver has not enough free memory it blocks the communication channel until enough memory is available. This receiver side blocking mechanism realizes the external back pressure behavior. The internal transport of data records between operators within a vertex follows a pull-based approach, where the root operators pulls individual records from the incoming transport buffer through all dependent operators and writes the results back to the outgoing transport buffer (see Fig. 9.4).

9.5.2 *Control-Flow Execution*

Control-flow operators can be seen as branching points between the execution of multiple partial physical plans. The control-flow tree resides centrally in the associated TC, where the evaluation of the current control-flow operation takes place. The processing of a dataflow program is carried out as a continuous interplay between

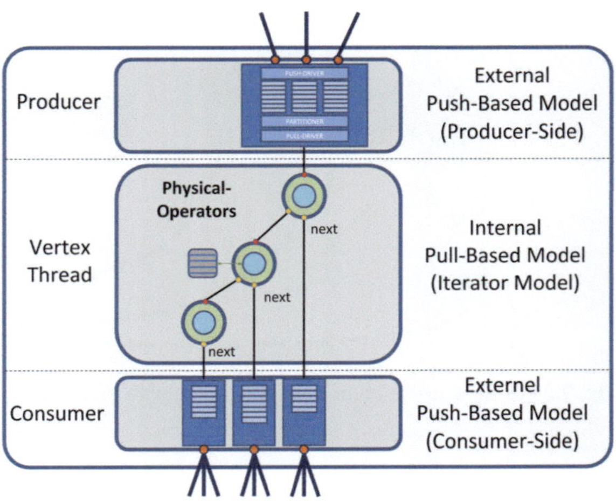

Fig. 9.4 Pull-based execution of data-parallel operators

the execution of partial physical plans and the stepwise interpretation of the centralized control-flow tree. If the execution of a partial physical plan is terminated, the results are materialized in the outgoing intermediate caches of the sink vertices. The results are kept in the cache until one or more subsequent partial physical plans make an request. Point of cache expiration, i.e. when no more intermediate results are required, can be determined via static analysis of the complete physical plan. The runtime coordination between control-flow evaluation and distributed execution is realized via a so-called *activation token* which is passed back and forth between the TMs and the TC.

9.6 Summary

A large class of programs require control flow primitives such as iterations and conditional branches. Support for control flow on the level of a dataflow engine eliminates the need to express these programs as multiple jobs on the client-side. Furthermore, it allows to optimize programs in their entirety, across what previously were separate jobs: more intermediate results can be kept in memory, while the selection and configuration of physical operators–including data partitioning–can take the entire program into account.

Aura, the novel parallel dataflow engine presented in this paper, integrates dedicated support for control flow with automatic plan compilation and an efficient distributed execution model. This design makes our prototype an interesting platform to research the optimal execution of advanced dataflow programs as, for example, from the domains of graph analysis and machine learning.

Acknowledgments This work has been supported through grants by the German Science Foundation (DFG) as FOR 1306 Stratosphere and by the German Ministry for Education and Research as Berlin Big Data Center BBDC (funding mark 01IS14013A).

References

1. Alexandrov A, Bergmann R, Ewen S, Freytag JC, Hueske F, Heise A, Kao O, Leich M, Leser U, Markl V, Naumann F, Peters M, Rheinlaender A, Sax MJ, Schelter S, Hoeger M, Tzoumas K, Warneke D (2014) The stratosphere platform for big data analytics. VLDB J 23(6):939–964
2. Alexandrov A, Kunft A, Katsifodimos A, Schüler F, Thamsen, L, Kao O, Herb T, Markl V (2015) Implicit parallelism through deep language embedding. In: Proceedings of the 2015 ACM SIGMOD international conference on management of data SIGMOD '15, ACM, New York
3. Battré D, Ewen S, Hueske F, Kao O, Markl V, Warneke D (2010) Nephele/PACTs: a programming model and execution framework for web-scale analytical processing. In: Proceedings of the 1st ACM symposium on cloud computing SoCC '10, ACM, New York
4. Chaiken R, Jenkins B, Larson PA, Ramsey B, Shakib D, Weaver, S, Zhou J (2008) SCOPE: easy and efficient parallel processing of massive data sets. Proc. VLDB Endow. 1(2) (2008)

5. Dean J, Ghemawat S (2004) MapReduce: simplified data processing on large clusters. In: Proceedings of the 6th conference on symposium on operating systems design & implementation. OSDI'04, USENIX Association
6. Ewen S, Tzoumas K, Kaufmann M, Markl V (2012) Spinning fast iterative data flows. Proc VLDB Endow 5(11):1268–1269
7. Isard M, Budiu M, Yu Y, Birrell A, Fetterly D (2007) Dryad: distributed data-parallel programs from sequential building blocks. In: Proceedings of the 2Nd ACM SIGOPS/EuroSys european conference on computer systems. EuroSys '07, ACM, New York
8. McSherry F, Murray DG, Isaacs R, Isard M (2013) Differential dataflow. In: Proceedings of the 6th conference on innovative data systems research (CIDR). CIDR'13, ACM, New York
9. Murray DG, McSherry F, Isaacs R, Isard M, Barham P, Abadi M (2013) Naiad: a Timely Dataflow System. In: Proceedings of the twenty-fourth ACM symposium on operating systems principles. ACM, New York
10. Olston C, Reed B, Srivastava U, Kumar R, Tomkins A (2008) Pig Latin: a not-so-foreign language for data processing. In: Proceedings of the 2008 ACM SIGMOD international conference on management of data. SIGMOD, ACM, New York
11. Shvachko K, Kuang H, Radia S, Chansler R (2010) The hadoop distributed file system. In: 2010 IEEE 26th symposium on mass storage systems and technologies (MSST)
12. Thusoo A, Sarma JS, Jain N, Shao Z, Chakka P, Anthony S, Liu H, Wyckoff P, Murthy R (2009) Hive: a warehousing solution over a map-reduce framework. VLDB
13. Warneke D, Kao O (2009) Nephele: efficient parallel data processing in the cloud. In: Proceedings of the 2Nd workshop on many-task computing on rids and supercomputers. MTAGS '09, ACM, New York
14. Yu Y, Isard M, Fetterly D, Budiu M, Erlingsson Ú, Gunda PK, Currey J (2008) DryadLINQ: a system for general-purpose distributed data-parallel computing using a high-level language. In: Proceedings of the OSDI'08: eighth symposium on operating system design and implementation. OSDI'08
15. Zaharia M, Chowdhury M, Das T, Dave A, Ma J, McCauley M, Franklin MJ, Shenker S, Stoica I (2012) Resilient distributed datasets: a fault-tolerant abstraction for in-memory cluster computing. In: Proceedings of the 9th USENIX conference on networked systems design and implementation. NSDI'12, USENIX Association
16. Zaharia M, Chowdhury M, Franklin MJ, Shenker S, Stoica I (2010) Spark: cluster computing with working sets. In: Proceedings of the 2Nd USENIX conference on hot topics in cloud computing. HotCloud'10, USENIX Association

Chapter 10
Parallel Code Analysis in HPC User Support

Rene Sitt, Alexandra Feith and Dörte C. Sternel

Abstract Over the last decades, the access to high performance computing resources has transitioned from being restricted to only a few specialists to becoming a commodity for people from widely varying contexts. Consequently, this includes the need for a level of in-depth user support which cannot be provided by a sysadmin alone. The Competence Center for High Performance Computing in Hesse aims to establish this service for the Hessian universities. As such, merely troubleshooting user scripts and codes is not enough; instead, the goal is to optimize the usage of computational resources, which needs to include the analysis of user codes and ensuring good performance and scalability. For this task, parallel code analysis tools in user support are becoming more and more commonplace and may indeed be regarded as essential. As a result, HPC user support specialists form a separate user group from application developers and require different sets of features.

10.1 Introduction

High Performance Computing (HPC), the ability to do massive computations on parallel systems, has become a useful research tool for a broad user base in the scientific community. This leads, in turn, to an increased need for user support in order to get the full potential out of the available HPC infrastructure. The Competence Center for High Performance Computing in Hesse (Hessisches Kompetenzzentrum für Hochleistungsrechnen, HKHLR) [7], aims to provide these services, and utilizes an array of parallel code performance analysis tools. Since the HKHLR staff members

R. Sitt (✉)
Hessisches Kompetenzzentrum für Hochleistungsrechnen, Hochschulrechenzentrum, Hans-Meerwein-Straße 6, 35032 Marburg, Germany
e-mail: sittr@hrz.uni-marburg.de

A. Feith · D.C. Sternel
Hessisches Kompetenzzentrum für Hochleistungsrechnen, Mornewegstraße 30, 64293 Darmstadt, Germany

A. Knüpfer et al. (eds.), *Tools for High Performance Computing 2015*,
DOI 10.1007/978-3-319-39589-0_10

themselves are no code developers per se (although being proficient with one or more of the common programming languages is a necessary prerequisite), their requirements on these tools are different from those of a "plain" software engineer.

10.2 The Competence Center for HPC in Hesse

The Competence Center for HPC in Hesse was founded by the universities of Darmstadt, Frankfurt/Main, Giessen, Marburg and Kassel in 2014 and is funded by the Hessen State Ministry of Higher Education, Research and the Arts. Each one of these universities maintains one or more computing clusters, two of which (the LICHTENBERG cluster in Darmstadt [9] and the LOEWE- CSC cluster in Frankfurt [4]) are considered "Tier 2" clusters, i.e. supraregional clusters, while the others—SKYLLA in Giessen [12], MARC2 in Marburg [8], and the LINUX-cluster in Kassel [10]—are Tier 3 (i.e. regional) clusters.

10.2.1 Structure and Goals

The overarching goal for this project is to establish unified structures and services for HPC users in Hesse, to increase the coordination and exchange of knowledge between the universities and to provide user support [6]. This entails holding introductory meetings for users who are new to HPC, day-to-day troubleshooting, organizing workshops, analyzing code performance, and supporting whole projects as well as providing, configuring, and porting necessary software (Fig. 10.1).

10.2.2 User Support in a HPC Environment

With HPC resources becoming more commonly available and accessible, the spectrum of potential users significantly broadens. In terms of familiarity with parallel

Fig. 10.1 Current HKHLR members

programming on computing clusters, three different types of users can be identified [6]:

Black box users with little or no previous background in HPC. Mostly, they will be using preinstalled software packages, which can be provided with a set of input data to produce a solution that is interpretable without knowledge about the actual computation. Main support issues in this case include the correct installation and configuration of the software package in question, as well as usage questions about the job scheduler and the cluster environment. From a performance optimization perspective, there is little to gain with correctly configured and well-written software aside from discovering bugs.

Source code augmenters, which are somewhat familiar with software programming in general, but not specifically in a massively parallel environment. Typically, most users of university computing clusters will fall in this category. Support issues include debugging, parallelization, and optimization of user-written programs, which requires at least a rough knowledge of programming languages like Java, R, Matlab, Python, and C. Performance-wise, this group has high potential for parallel code optimization.

Source code developers with specialized knowledge about parallel software development and optimization. This group is likely to want to perform code optimization for themselves, and user support will consist of mostly technical advice and providing access to tools for code analysis and debugging.

From a user support perspective, source code augmenters are, on the one hand, a group with high optimization potential, but often lacking motivation or expertise to do so. It is therefore an imperative for user supporters to actively get into contact with them and offer help with code analysis and—if needed—optimization. Members of the source code developers group, on the other hand, are likely concerned with their code performance from the beginning (and also likely to be more competent due to being more specialized than the supporter), so user support has primarily an assisting role.

10.3 Parallel Performance Analysis Tools: Requirements and User Insights

In order to reach the goals of the HKHLR, parallel code analysis tools have proven to be an invaluable and indeed essential asset.

10.3.1 Motivations for Parallel Analysis

In scientific computation—excluding developers of parallel software packages—, many users are primarily focused on getting results in an acceptable timeframe rather than investing additional work to make an acceptably performing code into an optimal performing one. This is understandable, since as long as the code itself is not the immediate subject of research, there might simply not be enough time for every individual scientist to do code optimization.

There are good reasons to optimize code; from a financial perspective, running a computing cluster is an expensive task and needs to produce tangible results in order to get funded. Thus, it is imperative to use the available capacities as efficiently as possible to justify investments for a high performance computing infrastructure. Furthermore, there are also ecological reasons to optimize the usage of cluster system ("Green IT"), as it is important to avoid wasting energy due to slow-running code or poor utilization of computing resources [1]. Lastly, performance optimization is also valuable from a scientific perspective, since fast and scalable code enables scientists to generate more results or data points in a given time and facilitates the computation of bigger and more complex systems than a poorly optimized code.

10.3.2 Requirements for Parallel Code Analysis Tools in HPC User Support

Although it may not be obvious at first, user supporters who utilize code analysis tools form a separate user group with different requirements than "plain" code developers. In general, there is a greater focus on accessibility and ease of use for the tools in question, since the roles of "code expert" (the person who wrote the code) and "parallelization expert" are most likely distributed between the user and the supporter, respectively, instead of a single expert developer.

First and foremost, it is important that everyone, both support staff and users, have access to these tools. This might seem trivial for open source programs, but the licensing model for non-free software has to fit within the conditions of a cluster environment.

Secondly, it is desirable that the usage of these tools does not generate too large work overheads, especially with respect to the amount of time needed to perform measurements until useful results can be obtained. Since user support usually does not allow for giving a single project exclusive focus over a long timespan, tools which generate performance data without relying on long runtimes or complex configurations are likely to be preferred.

In addition to that, using an analysis tool should be as intuitive as possible. In HPC user support the code experts and parallelization experts are different persons and performance optimization is likely to be carried out in a collaboration between both. A tool requiring extensive training to use effectively would not be feasible for this situation.

Lastly, combining the requirements of "little overhead" and "intuitive design", an ideal analysis tool should allow an experienced user to spot large performance issues (more or less) immediately.

10.3.3 Parallel Analysis Tools in Use at the HKHLR

At the HKHLR, there are currently several analysis tools in use, each serving a separate purpose. These include the SCALASCA suite, consisting of SCORE-P for profile and trace generation, CUBE [3] for profile file examination and SCALASCA [2] for trace analysis. Furthermore, VAMPIR [5, 13] is used for visual trace analysis. These tools fit together well, with SCORE- P-generated files readable by the other programs. Often, already a rough examination of a generated profile (with SCORE-P in textual and CUBE in visual form) gives first hints about possible performance bottlenecks, which can then be verified in analyzing the traces. The SCALASCA suite is freely available [11] while a VAMPIR state license was obtained some time ago, making it accessible for every member of the Hessian universities and satisfying the “accessibility” requirement mentioned in Sect. 10.3.2.

10.3.4 User Insights

In general, the tool setup described above works quite efficiently in a user support environment, as indicated in several successful parallel code optimizations (for a detailed case study, see [6]).

```
Estimated aggregate size of event trace:                    10GB
Estimated requirements for largest trace buffer (max_buf): 1319MB
Estimated memory requirements (SCOREP_TOTAL_MEMORY):        1321MB
(hint: When tracing set SCOREP_TOTAL_MEMORY=1321MB to avoid intermediate flushes
 or reduce requirements using USR regions filters.)

flt     type      max_buf[B]        visits time[s] time[%] time/visit[us]  region
         ALL 1,382,076,273 349,184,300 8550.08   100.0           24.49  ALL
         USR   805,308,384 268,436,128  402.08     4.7            1.50  USR
         MPI   576,767,073  80,747,900 1980.29    23.2           24.52  MPI
         COM           816         272 6167.70    72.1     22675380.81  COM

         USR   402,653,184 134,217,728   59.66     0.7            0.44  V
         USR   402,653,184 134,217,728   50.10     0.6            0.37  int volatile
         MPI   288,358,600  40,370,190  166.74     2.0            4.13  MPI_Recv
         MPI   288,358,500  40,370,190   54.83     0.6            1.36  MPI_Send
         MPI        29,792       3,136  450.27     5.3       143580.41  MPI_Sendrecv
         MPI        10,488       1,472  853.79    10.0       580017.65  MPI_Allreduce
         MPI         4,344       1,448    0.00     0.0            0.52  MPI_Comm_size
         MPI         1,248         416    0.00     0.0            2.76  MPI_Comm_free
         MPI         1,200         400    0.03     0.0           72.80  MPI_Comm_dup
         MPI         1,152         384    0.00     0.0            0.82  MPI_Comm_rank
         MPI         1,026         144  449.01     5.3      3118122.20  MPI_Bcast
         COM           792         264  163.68     1.9       619987.27  curr
         USR           792         264    0.05     0.0          195.13  fx
         MPI           627          88    0.44     0.0         4957.23  MPI_Reduce
         USR           504         168    0.00     0.0            7.54  field
         USR           480         160  197.21     2.3      1232567.11  op_ev_pot
         USR           240          80   95.06     1.1      1188207.03  op_ev_kin
         MPI            48          16    0.00     0.0           71.33  MPI_Comm_split
         MPI            24           8    2.21     0.0       276192.75  MPI_Init
         MPI            24           8    2.98     0.0       372657.00  MPI_Finalize
         COM            24           8 6004.03    70.2    750503367.80  main
```

Fig. 10.2 Example of an unoptimized code leading to a large trace file size with only 8 MPI processes

A persistent and frequent problem is the enormous size of any trace file with process counts larger than 10–20 or with individual functions being called millions of times; in fact, the latter often coincides with an unoptimized code.

For example, the trace of a physics simulation with a poor communication model can get quite large with only 8 MPI processes running (see Fig. 10.2). The profile clearly displays the unusually large amount of MPI calls, which warrants a closer inspection, but also results in an unwieldy trace file.

Of course, SCORE-P offers to record traces in a compressed format to mitigate this problem; however, the possibility of recording traces for very complex code and a very large amount of processes is still limited by the available resources (e.g. memory and disk space).

The visualization of such large traces can lead to the GUI behaving sluggishly, depending on available system resources. There is, of course, an option to start a multi-process analysis task (utilizing VAMPIRSERVER), but with high load on the cluster (e.g. the percentage of occupied cores on the MaRC2 cluster in Marburg reaches 90 % and higher quite often), this might imply additional waiting time until the server task can start.

In addition, it was possible to reduce the amount of MPI calls from the code (see Fig. 10.2), leading to a reduced trace footprint (which can be reduced further by filtering out the most-visited USR functions), as can be seen in Fig. 10.3.

```
Estimated aggregate size of event trace:                   7GB
Estimated requirements for largest trace buffer (max_buf): 769MB
Estimated memory requirements (SCOREP_TOTAL_MEMORY):       771MB
(hint: When tracing set SCOREP_TOTAL_MEMORY=771MB to avoid intermediate flushes
 or reduce requirements using USR regions filters.)

flt     type max_buf[B]       visits time[s] time[%] time/visit[us]  region
         ALL 805,343,543 268,441,056 1163.35   100.0           4.33  ALL
         USR 805,308,384 268,436,128  411.81    35.4           1.53  USR
         MPI      34,343       4,656  267.24    23.0       57397.77  MPI
         COM         816         272  484.29    41.6     1780483.58  COM

         USR 402,653,184 134,217,728   58.40     5.0           0.44  V
         USR 402,653,184 134,217,728   50.46     4.3           0.38  int volatile
         MPI      21,280       2,240  145.79    12.5       65086.45  MPI_Sendrecv
         MPI       5,472         768    0.02     0.0          19.91  MPI_Allreduce
         MPI       1,368         192  111.66     9.6      581558.15  MPI_Bcast
         MPI       1,300         168    0.92     0.1        5454.66  MPI_Recv
         MPI       1,200         168    5.21     0.4       30985.08  MPI_Send
         MPI       1,176         392    0.00     0.0           0.57  MPI_Comm_size
         USR         792         264    0.03     0.0          95.28  fx
         COM         792         264   73.87     6.3      279792.91  curr
         MPI         672         224    0.00     0.0           1.69  MPI_Comm_free
         MPI         627          88    0.15     0.0        1741.92  MPI_Reduce
         MPI         624         208    0.18     0.0         877.88  MPI_Comm_dup
         MPI         528         176    0.00     0.0           0.75  MPI_Comm_rank
         USR         504         168    0.00     0.0           4.87  field
         USR         480         160  203.99    17.5     1274955.35  op_ev_pot
         USR         240          80   98.93     8.5     1236590.68  op_ev_kin
         MPI          48          16    0.00     0.0          70.67  MPI_Comm_split
         MPI          24           8    0.45     0.0       56027.00  MPI_Finalize
         MPI          24           8    2.87     0.2      358508.68  MPI_Init
         COM          24           8  410.43    35.3    51303275.62  main
```

Fig. 10.3 Profile of the same code as in Fig. 10.2, after an optimization of the communication model, greatly reducing the frequency of point-to-point MPI calls

Another virtually unavoidable problem arises from having different compilers and MPI implementations on a cluster to allow users to choose their desired programming environment freely. With respect to parallel analysis tools, this means that maintaining parallel installations of these tools to fit different compiler/MPI-combinations is a must, which can get unwieldy. For example, the MARC2 cluster in Marburg has several versions of `gcc`, `icc` and `pgcc` installed, as well as both `OpenMPI` and `Parastation MPI`, leading to—theoretically—six installations of SCORE-P and SCALASCA to cover every combination.

10.4 Conclusion

In summary, the need for user support in high performance computing is rising with the demand for and growing accessibility of clustered computational resources. With this field opening up for non-specialists and non-programmers, user support can no longer consist of troubleshooting only; it needs to offer active supervision of projects, performance monitoring, and code optimization. In consequence, this makes parallel analysis essential. With the tools currently in use, performance analysis has already become a much less daunting task than it used to be in the past.

References

1. Bischof, C., an Mey, D., Iwainsky, C.: Brainware for green HPC. Computer Science - Research and Development. Springer, New York (2011). doi:10.1007/s00450-011-0198-5
2. Geimer, M., Wolf, F., Wylie, B.J.N., Ábrahám, E., Becker, D., Mohr, B.: The Scalasca performance toolset architecture. Concurr. Comput.: Pract. Exp. **22**, 702–719 (2010)
3. Geimer, M., Saviankou, P., Strube, A., Szebenyi, Z., Wolf, F., Wylie, B.J.N.: Further improving the scalability of the Scalasca toolset. In: Proceedings of PARA, Reykjavik, Iceland (2012)
4. http://csc.uni-frankfurt.de
5. Knüpfer, A., Brunst, H., Doleschal, J., Jurenz, M., Lieber, M., Mickler, H., Müller, M.S., Nagel, W.E.: The vampir performance analysis tool-set. In: Resch, M., Keller, R., Himmler, V., Krammer, B., Schulz, A. (eds.) Tools for High Performance Computing, pp. 139–155. Springer, Heidelberg (2008). doi:10.1007/978-3-540-68564-7
6. Sternel, D.C., Iwainsky, C., Opfer, T., Feith, A.: The Hessian competence centre for high performance computing: "Brainware" for Green IT. In: Ivnyi, P., Topping, B.H.V. (eds.) Proceedings of the Fourth International Conference on Parallel, Distributed, Grid and Cloud Computing for Engineering. Civil-Comp Press, Stirlingshire, UK, Paper 37 (2015). doi:10.4203/ccp.107.37
7. www.hpc-hessen.de
8. www.uni-marburg.de/hrz/infrastruktur/zserv/cluster
9. www.hhlr.tu-darmstadt.de
10. www.uni-kassel.de/its-handbuch/daten-dienste/wissenschaftliche-datenverarbeitung.html
11. www.scalasca.de
12. www.uni-giessen.de/cms/fbz/svc/hrz/svc/server/hpc
13. www.vampir.eu

Chapter 11
PARCOACH Extension for Hybrid Applications with Interprocedural Analysis

Emmanuelle Saillard, Hugo Brunie, Patrick Carribault and Denis Barthou

Abstract Supercomputers are rapidly evolving with now millions of processing units, posing the questions of their programmability. Despite the emergence of more widespread and functional programming models, developing correct and effective parallel applications still remains a complex task. Although debugging solutions have emerged to address this issue, they often come with restrictions. Furthermore, programming model evolutions stress the requirement for a validation tool able to handle hybrid applications. Indeed, as current scientific applications mainly rely on MPI (Message-Passing Interface), new hardwares designed with a larger node-level parallelism advocate for an MPI+X solution with X a shared-memory model like OpenMP. But integrating two different approaches inside the same application can be error-prone leading to complex bugs. In an MPI+X program, not only the correctness of MPI should be ensured but also its interactions with the multi-threaded model. For example, identical MPI collective operations cannot be performed by multiple non-synchronized threads. In this paper, we present an extension of the PARallel COntrol flow Anomaly CHecker (PARCOACH) to enable verification of hybrid HPC applications. Relying on a GCC plugin that combines static and dynamic analysis, the first pass statically verifies the thread level required by an MPI+OpenMP application and outlines execution paths leading to potential deadlocks. Based on this analysis, the code is selectively instrumented, displaying an error and interrupting all processes if the actual scheduling leads to a deadlock situation.

11.1 Introduction

The evolution of supercomputers to Exascale systems raises the issue of choosing the right parallel programming models for applications. Currently, most HPC applications are based on MPI. But the hardware evolution of increasing core counts

E. Saillard (✉) · H. Brunie · P. Carribault
CEA, DAM, DIF, 91297 Arpajon, France
e-mail: emmanuelle.saillard.ocre@cea.fr

D. Barthou
Bordeaux Institute of Technology, LaBRI / INRIA, Bordeaux, France

A. Knüpfer et al. (eds.), *Tools for High Performance Computing 2015*,
DOI 10.1007/978-3-319-39589-0_11

per node leads to a mix of MPI with shared-memory approaches like OpenMP. However merging two parallel programming models within the same application requires full interoperability between these models and makes the debugging task more challenging. Therefore, there is a need for tools able to identify functional bugs as early as possible during the development cycle. To tackle this issue, we designed the PARallel COntrol flow Anomaly CHecker (PARCOACH) that combines static and dynamic analyses to enable an early detection of bugs in parallel applications. With the help of a compiler pass, PARCOACH can extract potential parallel deadlocks related to control-flow divergence and issue warnings during the compilation. Not only the parallel constructs involved in the deadlock are identified and printed during the compilation, but the statements responsible for the control-flow divergence are also outputted. In this paper, we propose an extension of PARCOACH to hybrid MPI+OpenMP applications and an interprocedural analysis to improve the bug detection through a whole program. This work is based on [10] and extends [11] with more details and an interprocedural analysis. To the best of our knowledge, only Marmot [3] is able to detect errors in MPI+OpenMP programs. But as a dynamic tool, Marmot detects errors during the execution and is limited to the dynamic parallel schedule and only detects errors occurring for a given inputset whereas our approach allows for static bug detection with runtime support and detects bugs for all possible values of inputs.

In the following we assume SPMD MPI programs that call all MPI collective operations with compatible arguments (only the `MPI_COMM_WORLD` communicator is supported). Therefore, each MPI task can have a different control flow within functions, but it goes through the same functions for communications. Issues related to MPI arguments can be tested through other tools.

11.1.1 Motivating Examples

The MPI specification requires that all MPI processes call the same collective operations (blocking and non-blocking since MPI-3) in the same order [6]. These calls do not have to occur at the same line of source code, but the dynamic sequence of collectives should be the same otherwise a deadlock can occur. In addition, MPI calls should be cautiously located in multi-threaded regions. Focusing only on MPI, in Listing 1, because of the conditional in line 2 (`if` statement), some processes may call the `MPI_Reduce` function while others may not. Similarly, in Listing 2, some MPI ranks may perform a blocking barrier (`MPI_Barrier`) while others will call a non-blocking one (`MPI_Ibarrier`). The sequence is the same (call to one barrier), but this blocking/non-blocking matching is forbidden by the MPI specification (Fig. 11.1).

Regarding hybrid MPI+OpenMP applications, the MPI API defines four levels of thread support to indicate how threads should interact with MPI: `MPI_THREAD_SINGLE`, `MPI_THREAD_FUNNELED`, `MPI_THREAD_SERIALIZED` and `MPI_THREAD_MULTIPLE`. MPI processes can be multithreaded but

Listing 11.1

```
void f(){
  if (...)
  {
     #pragma omp parallel
     {
         #pragma omp single
         {
             MPI_Reduce(..)
         }
     }
  }
}
```

Listing 11.2

```
void f(){
  if (...)
     MPI_Barrier(..)
  else
     MPI_Ibarrier(...)
  #pragma omp parallel
  {
    /***/
  }
}
```

Listing 11.3

```
1  void f(){
2    #pragma omp parallel
3    {
4      /***/
5      #pragma omp master
6      {
7         MPI_Send(..)
8         /***/
9      }
10   }
11 }
```

Listing 11.4

```
void f(){
  #pragma omp parallel
  {
    #pragma omp single nowait
    {
      MPI_Reduce(..)
    }
    #pragma omp single
    {
      MPI_Reduce(..)
    }
  }
}
```

Fig. 11.1 MPI+OpenMP Examples with different uses of MPI calls

the MPI standard specifies that "it is the user responsibility to prevent races when threads within the same application post conflicting communication calls" [6]. In Listing 2, MPI calls are executed outside the multithreaded region. This piece of code is therefore compliant with the `MPI_THREAD_SINGLE` level. But MPI communications may appear inside OpenMP blocks. For example, the MPI point-to-point function at line 7 in Listing 3 is inside a `master` block. The minimum thread level required for this code is therefore `MPI_THREAD_FUNNELED`. However, calls located inside a single or master block may lead to different thread support. Indeed, in Listing 4, two `MPI_Reduce` are in different `single` regions. Because of the `nowait` clause on the first `single` region, these calls are performed simultaneously by different threads. This example requires the maximum thread support level i.e., `MPI_THREAD_MULTIPLE`.

These simple examples illustrate the difficulty for a developer to ensure that MPI calls are correctly used inside an hybrid MPI+OpenMP application. A tool able to check, for each MPI call, in which thread context it can be performed would help the application developer to know which thread-level an application requires. Furthermore, beyond this support, checking deadlock of MPI collective communications in presence of OpenMP constructs can be very tricky. In this paper, we propose an extension of PARCOACH to tackle these issues, with the help of an interprocedural analysis to improve the compile-time detection.

Section 11.2 gives an overview of the PARCOACH platform with a description of its static and dynamic analyses for hybrid MPI+OpenMP applications. Then, Sect. 11.3 describes an interprocedural extension of the PARCOACH static pass. Section 11.4 presents experimental results and finally Sect. 11.5 concludes the paper.

11.2 PARCOACH Static and Dynamic Analyses for Hybrid Applications

PARCOACH uses a two-step method to verify MPI+OpenMP applications as shown in Fig. 11.2. The first analysis is located in the middle of the compilation chain, where the code is represented as an intermediate form. Each function of a program is depicted by a graph representation called *Control Flow Graph* (CFG). PARCOACH analyses the CFG of each function to detect potential errors or deadlocks in a program. When a potential deadlock is detected, PARCOACH reports a warning with precise information about the possible deadlock (line and name of the guilty MPI communications, and line of conditionals responsible for the deadlock). Then the warnings are confirmed by a static instrumentation of the code. Note that whenever the compile-time analysis is able to statically prove the correctness of a function, no code is inserted in the program, reducing the impact of our transformation on the execution time. If deadlocks are about to occur at runtime, the program is stopped and PARCOACH returns error messages with compilation information.

This section describes the following new features of PARCOACH: (i) detection of the minimal MPI thread-level support required by an MPI+OpenMP application (see [10] for more details) and (ii) checking misuse of MPI blocking and nonblocking collectives in a multi-threaded context (extension of [11]).

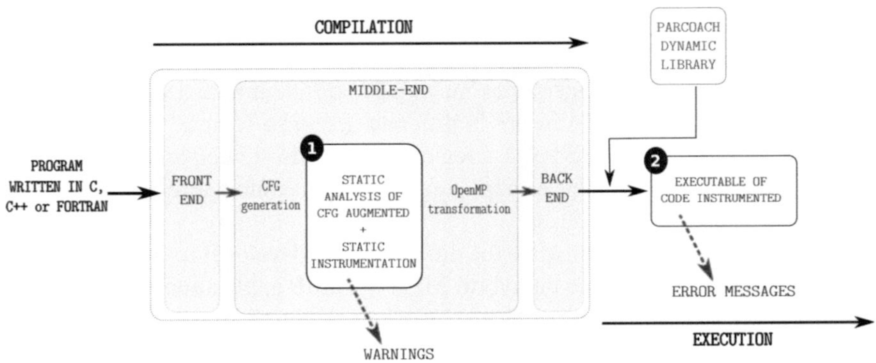

Fig. 11.2 PARCOACH two-step analysis overview

11.2.1 MPI Thread-Level Checking

This analysis finds the right MPI thread-level support to be used and identifies code fragments that may prevent conformance to a given level. Verifying the compliance of an MPI thread level in MPI+OpenMP code resorts to check the placement of MPI calls. To determine the thread context in which MPI calls are performed, we augment the CFGs by marking the nodes containing MPI calls (point-to-point and collective). Then, with a depth-first search traversal, we associate a *parallelism word* to each node. As defined in [10], a parallelism word is the sequence of OpenMP parallel constructs (P:`parallel`, S:`single`, M:`master` and B:`barrier` for implicit and explicit barriers) surrounding a node from the beginning of the function to the node. The analysis detects CFG nodes containing MPI calls associated to parallelism words defining a multithreaded context and forbidden concurrent calls. Based on this analysis, the following section describes how collectives operations can be verified in a multithreaded context.

11.2.2 MPI Collective Communication Verification

This analysis proposes a solution to check the sequence of collective communications inside MPI+OpenMP programs. PARCOACH verifies that there is a total order between the MPI collective calls within each process and it ensures that this order is the same for all MPI ranks. Our analysis relies on checking 3 rules:

1. Within an MPI process, all collectives are executed in a monothreaded context;
2. Within an MPI process, two collective executions are sequentially ordered, either because they belong to the same monothreaded region or because they are separated by a thread synchronization (no concurrent monothreaded regions);
3. The sequence of collectives are the same for all MPI processes (i.e., sequences do not depend on the control flow).

A function is then said to be potentially statically incorrect if at least one of the three categories presented in Fig. 11.3 is verified. This section describes how these error categories can be detected.

Category 1 Detection: This phase of the static analysis corresponds to the detection of MPI collectives that are not executed in a monothreaded region. To this end, we use the parallelism words defined in [11]. A parallelism word defines a monothreaded context if it ends with an S or an M (ignoring Bs). If the parallelism word has a sequence of two or more P with no S or M in-between, it implies the parallelism is nested. Even if the word ends with an S or M, one thread for each thread team can execute the MPI collectives.

For this part, it is not necessary to separate `single` from `master` regions. So the finite-state automaton in [10] is simplified into the automaton presented Fig. 11.4. It recognizes the language of parallelism words corresponding to monothreaded

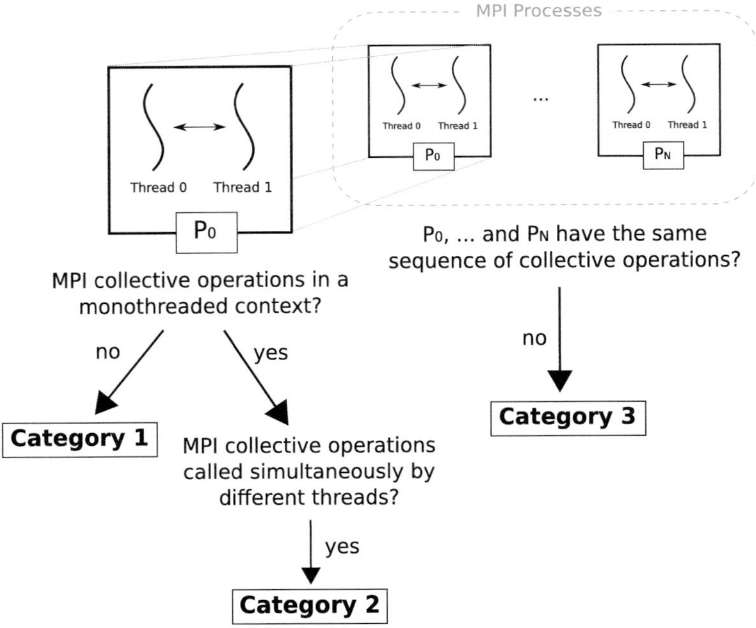

Fig. 11.3 Categories of possible errors in a hybrid program with N MPI processes and two threads per process

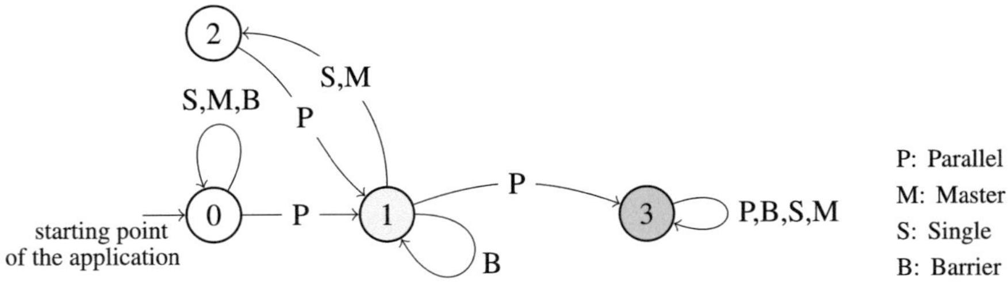

Fig. 11.4 Automata of possible parallelism words. Nodes 0 and 2 correspond to code executed by the master thread or a single thread. Node 1 corresponds to code executed in a parallel region, and 3 to code executed in nested parallel region

regions. States 0 and 2 are the accepting states and the language L defined by $L = (S|M|B|PB*S|PB*M)*$ contains the accepted words (parallelism words ending by S or M without a repeated sequence of P).

Category 2 Detection: For this analysis, MPI collective operations are assumed to be called in monothreaded regions, as defined in the previous section. However, different MPI collectives can still be executed simultaneously if monothreaded regions are executed in parallel. This phase corresponds to the detection of MPI collective calls in concurrent monothreaded regions.

Algorithm 1 Step 1: Static Pass of hybrid programs

function HYBRID_STATIC_PASS($G = (V, E), L$)
  ▷ G: CFG, L: language of correct parallelism words
  DFS(G, $entry(G)$) ▷ parallelism words construction
  MULTITHREADED_REGIONS(G, L) ▷ creates set S_{ipw}
  CONCURRENT_CALLS(G) ▷ creates set S_{cc}
  STATIC_PASS(G) ▷ creates set S
end function

Two nodes n_1 and n_2 are said to be in concurrent monothreaded regions if they are in monothreaded regions and if their parallelism words $pw[n_1]$ and $pw[n_2]$ are respectively equal to $wS^j u$ and $wS^k v$ where w is a common prefix (possibly empty) with $j \neq k$, u and v words in $(P|S|B)*$.

Category 3 Detection: Once the sequence of MPI collective calls are verified in each MPI process, we must check that all sequences are the same for all processes. To verify that we rely on Algorithm 1 proposed in [9] with the extension of non-blocking collectives detailed in [4]. It detects MPI blocking and non-blocking collective mismatches by identifying conditionals potentially leading to a deadlock situation (set S). A warning is also issued for collective calls located in a loop as they can be called different times if the number of iterations is not the same for all MPI processes.

Static Pass Algorithm: To wrap-up all static algorithms, Algorithm 1 shows how analyses are combined. First the *DFS* function creates parallelism words. Then *MULTITHREADED_REGIONS* and *CONCURRENT_CALLS* procedures respectively detect categories 1 and 2 of errors. Finally the *STATIC_PASS* procedure detects category 3 of errors.

11.2.2.1 Static Instrumentation

The compile-time verification outputs warnings for MPI collective operations that may lead to an error or deadlock. Nevertheless the static analysis could lead to false positives if the actual control-flow divergence is not happening during the execution. To deal with this issue, we present a dynamic instrumentation that verifies warnings emitted at compile-time.

To dynamically verify the total order of MPI collective sequences in each MPI process, validation functions (CCipw and CCcc) are inserted in nodes in the sets Sipw and Scc generated by the static pass (see Algorithm 1). These functions are depicted in Algorithm 2. Function CCipw detects incorrect execution parallelism words and Function CCcc detects concurrent collective calls. To dynamically verify the total order of MPI collective sequences between processes, a check collective function *CC* is inserted before each MPI collective operation and before `return` statements. *CC* is depicted in Algorithm 2 in [8]. It takes as input the communicator com_c related to the collective call c and a color i_c specific to the type of collective. As multiple threads may call *CC* before `return` statements, this function is wrapped

Algorithm 2 Library Functions To Check MPI collectives

function CC_{ipw} ▷ Detect collectives in multithreaded regions
  if $pw_e \notin L_e$ **then**
    MPI_ABORT(com, 0)
  end if
end function

function CC_{cc} ▷ Detect concurrent collective calls
  CC_{ipw}
  if $collective\_lock = 1$ **then**
    MPI_ABORT(com, 0)
  else
    #pragma omp atomic write
    $collective\_lock = 1$
  end if
end function

function $CC(com_c, i_c)$ ▷ Detect collective calls mismatches

Algorithm 3 Step 2: Selective Static Instrumentation

function INSTRUMENTATION($communicator, G, S, S_{ipw}, S_{cc}$)
  ▷ G: CFG, S, S_{ipw}, S_{cc}: sets created at compile-time
  if $S \cup S_{ipw} \cup S_{cc} \neq \emptyset$ **then**
    ** **STEP 1: Control flow errors detection** **
    for n in nodes containing a call to collective c **do**
      Insert call to CC(com_c, i_c) before the call to c
    end for
    Before `return` statements insert
      `# pragma omp single`
        `CC(`$communicator$, 0`)}`
    ** **STEP 2: Collectives in multithreaded regions** **
    for $n \in S_{ipw}$ **do**
      Insert call to $CC_{ipw}()$ as the first statement of n
    end for
    ** **STEP 3: Concurrent MPI calls detection** **
    for $n \in S_{cc}$ **do**
      Insert call to $CC_{cc}()$ as the first statement of n
      Insert $collective\_lock = 0$ after the barrier(s) successors of the region created by n
    end for
  end if
end function

into a `single` pragma. Each function of a program is instrumented by Algorithm 3. If an error is about to occur, the program is stopped and an error message is returned with error type information.

11.3 Interprocedural Analysis

Because PARCOACH relies on an intraprocedural analysis, it miss errors across function boundaries and therefore it may produce false positive as well as false negative results. To extend PARCOACH with an interprocedural mechanism, we extended the intraprocedural approach through the application Call Graph (CG): nodes represent functions and edges model possible calls.

The main idea is to compute and reuse the summaries of each CFG through a CG traversal in reverse invocation order. For this purpose, the intraprocedural analysis is modified to return the valid sequence of collective operations for each function ($ValidSeq$). Based on this sequence of collective operations that all MPI processes will encounter, Algorithm 4 presents the interprocedural analysis. It takes as input the CG of a program and all the CFGs and returns the set O of conditionals that can lead to a deadlock.

Algorithm 4 Interprocedural Analysis

function INTERPROCEDURAL_ANALYSIS($\bigcup_f CFG_f, CG$) ▷ CG: Callgraph
  $Seq \leftarrow \{\}, O \leftarrow \emptyset$
  for each $n \in CG$ in reverse topological order **do**
    $n.ValidSeq \leftarrow \{\}$, $O_n \leftarrow \emptyset$
    for each $f \in SUCC_{CG}(n)$ **do**
      Replace f in n by $f.ValidSeq$
    end for
    $(O_n, ValidSeq) \leftarrow$ INTRAPROCEDURAL_ANALYSIS(CFG_n)
    $n.ValidSeq \leftarrow ValidSeq, O \leftarrow O \cup O_n$
  end for
  return O
end function

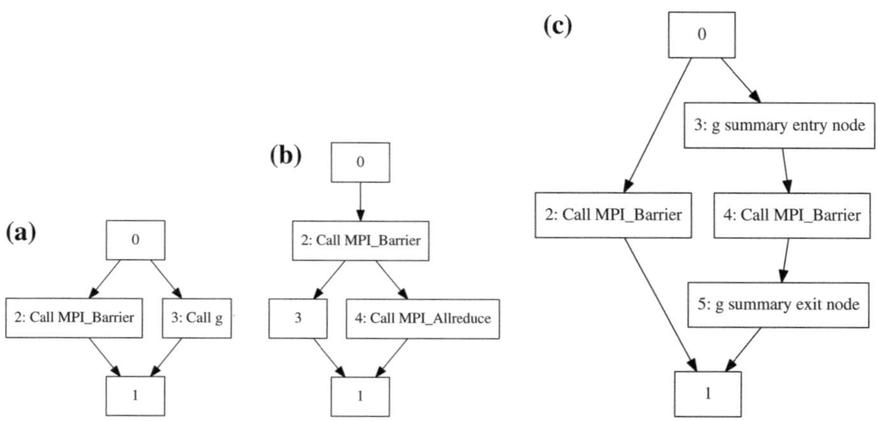

Fig. 11.5 Example of interprocedural analysis. **a** CFG of main. **b** CFG of g. **c** Program CFG

Figure 11.5a shows the main CFG calling function g (whose CFG is depicted in Fig. 11.5b). Performing our intraprocedural analysis would lead to a deadlock warning on main (for the MPI_Barrier operations) and on g (for the MPI_Allreduce collective). The resulting CFG after applying Algorithm 4 is illustrated in Fig. 11.5c: the call to g is replaced by the sequence of collective executed in g. Then, simply invoking the intraprocedural analysis on this new CFG results in no warning because there is one call to MPI_Barrier on each path of the program. Therefore, the only warning issued by our combined analysis is related to the call to MPI_Allreduce in function g (related to the if statement in node 2).

11.4 Experimental Results

We extended the PARCOACH implementation (GCC 4.7 plugin) to add analysis of hybrid applications. Thus, it is associated to GCC but simple to deploy in existing environments as it does not modify the compilation chain. To show the impact of PARCOACH analysis on the compilation and execution time, we tested the NAS-MZ [7], AMG benchmark [2], the EPCC suite [1] and HERA [5]. All results were conducted on Tera100, a petaflopic supercomputer at the CEA.

Figure 11.6 displays the overhead of compiling the applications with PARCOACH (only with our static analysis, or with the analysis and the static instrumentation). It shows that PARCOACH introduces a low compilation overhead (under 6 %). The execution time overheads obtained for the NAS benchmarks and HERA are presented in Fig. 11.7 running on MPICH with GCC OpenMP. The overheads obtained are under 25 % which is reasonable for debugging purpose.

The interprocedural analysis has also been implemented and integrated in PARCOACH (as a GCC plugin combined with a Python script). Figure 11.8 shows pieces of code from the IO module of the CORAL benchmark HACC. The calls

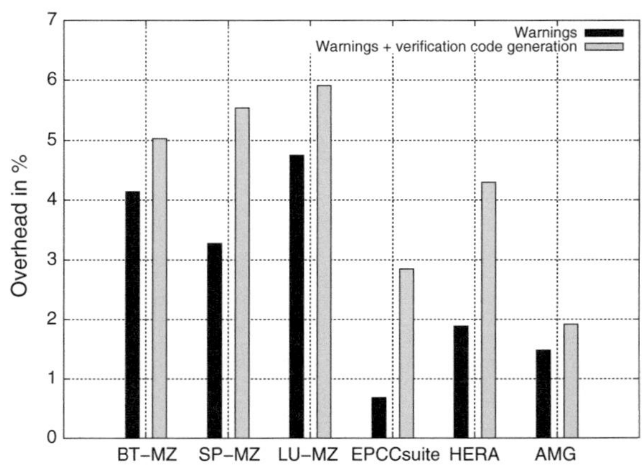

Fig. 11.6 Overhead of average compilation time with and without verification code generation

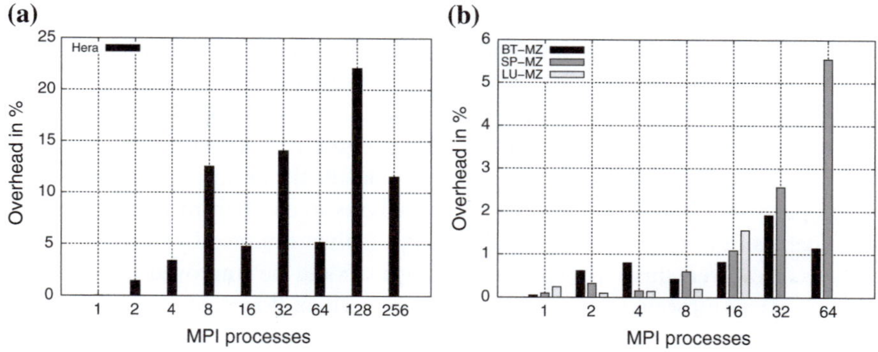

Fig. 11.7 Execution-Time Overhead for MZ (NAS class B) and HERA with 8 threads per MPI process (Strong scaling). **a** Execution-Time Overhead for HERA. **b** Execution-Time Overhead for NASPB-MZ

(a)
```
int RestartIO_GLEAN :: Close (void) {
    int status = -1;
    if (m_mode == WRITE_CHECKPOINT)
    {
        switch (m_interface)
        {
            case USE_POSIX:
                status = this->__POSIX_Close_Checkpoint(); break;
            case USE_MPIIO:
                status = this->__MPIIO_Close_Checkpoint(); break;
        }
    }
    else if (m_mode == READ_RESTART)
```

(b)
```
int RestartIO_GLEAN :: __MPIIO_Close_Checkpoint (void)
{
    [...]
    MPI_Barrier(m_partitionComm);
}
```

(c)
```
int RestartIO_GLEAN :: __POSIX_Close_Checkpoint (void)
{
    [...]
    MPI_Barrier(m_partitionComm);
    [...]
    MPI_Barrier(m_partitionComm);
    [...]
}
```

Fig. 11.8 Pieces of HACC/IO module code

in Fig. 11.8b, c contain one and two calls to an MPI collective, respectively. Hence when the function in Fig. 11.8a calls the others in different paths because of the switch, the execution could deadlock if the processes follow different paths. Of course the conditional statement does not depend on the rank number of each process, and therefore this is just a false positive. This interprocedural must be extended to a data flow analysis with the aim to study the dependence of these condition variables in order to know if they depend on the process rank or not.

11.5 Conclusion

The MPI+OpenMP approach is one solution to tackle the increasing node-level parallelism and the decreasing amount of memory per compute unit. Some production codes are already hybrid and other applications are in the development process. It is driven by available tools that could help debugging. That is why we developed the platform PARCOACH that helps application developers to check which interaction support is required for a specific hybrid code and checks the correct usage of blocking and non-blocking MPI collective communications in an MPI+OpenMP application. The main advantage of PARCOACH is that it highlights the statements responsible for the execution path potentially leading to future deadlocks or unspecified behaviors. We propose an adaptation of PARCOACH analyses to an interprocedural analysis. This enables us to reduce the number of false positives returned by the initial static analysis. However, this interprocedural analysis could be improved to propagate collective issue information and can be coupled to a data-flow analysis to avoid false positive results.

References

1. Bull JM, Enright JP, Guo X, Maynard C, Reid F (2010) Performance evaluation of mixed-mode openmp/mpi implementations. Intl Jf Parallel Program 38(5–6): 396–417
2. CORAL Benchmarks. https://asc.llnl.gov/CORAL-benchmarks/
3. Hilbrich T, Müller MS, Krammer B (2008) Detection of violations to the MPI standard in Hybrid open MP/MPI applications. In: International conference on OpenMP in a new era of parallelism, Springer, Heidelberg, pp 26–35
4. Jaeger J, Saillard E, Carribault P, Barthou D (2015) Correctness analysis of MPI-3 Non-blocking communications in PARCOACH. In: Proceedings of the 22Nd European MPI Users' group meeting, EuroMPI '15, ACM, New York, pp 16:1–16:2
5. Jourdren H (2003) HERA: A hydrodynamic AMR Platform for multi-physics simulations. In: Plewa T, Linde T, Weirs VG (eds) Adaptive mesh refinement - theory and applications. Springer, Heidelberg, pp 283–294
6. Message Passing Interface Forum. MPI: A Message-Passing Interface Standard Version 3.1, June 2015. http://www.mpi-forum.org/docs/mpi-3.1/mpi31-report.pdf
7. NASPB site: http://www.nas.nasa.gov/software/NPB
8. Saillard E, Carribault P, Barthou D (2013) Combining static and dynamic validation of mpi collective communications. In: Proceedings of the European MPI Users' group meeting. EuroMPI'13, ACM, New York, pp 117–122
9. Saillard E, Carribault P, Barthou D (2014) PARCOACH: Combining static and dynamic validation of MPI collective communications. Int J High Perform Comput Appl
10. Saillard E, Carribault P, Barthou D (2015) MPI Thread-level checking for MPI+OpenMP applications. In: EuroPar
11. Saillard E, Carribault P, Barthou D (2015) Static/dynamic validation of MPI collective communications in multi-threaded context. In: Proceedings of the 20th ACM SIGPLAN Symposium on principles and practice of parallel programming. PPoPP 2015, ACM, New York, pp 279–280

Chapter 12
Enabling Model-Centric Debugging for Task-Based Programming Models—A Tasking Control Interface

Mathias Nachtmann and José Gracia

Abstract This paper introduces the Tasking Control API—an API which allows tools, e.g. debuggers, to control the execution of task-based programming models as for instance OpenMP or OmpSs. However, the design of Tasking Control API is sufficiently generic to be applied to any other programming model for asynchronous task parallelization. We call this debugging approach model-centric debugging as opposed to language-centric debugging. The specification of Tasking Control API follows the spirit of the OpenMP Tools Interface to allow simple interoperability or even a future integration of Tasking Control API into the OpenMP Tools Interface.

12.1 Introduction

Debugging is a significant part of any software development process. However, what do we mean when we say debugger and debugging? Most practical definitions of debugging would probably include the following three aspects: (1) the ability to control execution of a program, in particular the ability to suspend and resume the program execution, (2) inspect the program state, e.g. print a variable's current value, and optionally (3) change the program state, e.g. set a variable's value or set the point where execution will be resumed.

Traditionally, debuggers are specific to a programming language. For instance, pydb is a debugger for Python, while gdb is a debugger for C (and C++, Fortran, and many others). All of these allow to reason about program at the lowest level that is accessible by the programming language, for instance static variable, pointers and function in the case of C. Today however, the programming language is just a small part of the programmer's development environment: we rely on third-party libraries, frameworks or whole programming models. In the ideal case, debuggers are *aware* of the development environment as a whole, operate on suitable *abstraction* layers, use the appropriate *vocabulary*, and *interact* with the runtime system. We call this *model-centric* [12] debugging as opposed to *language-centric* debugging.

M. Nachtmann (✉) · J. Gracia
High Performance Computing Center Stuttgart (HLRS), Stuttgart, Germany
e-mail: nachtmann@hlrs.de

A. Knüpfer et al. (eds.), *Tools for High Performance Computing 2015*,
DOI 10.1007/978-3-319-39589-0_12

In high-performance computing (HPC), the standard programming model for shared-memory systems is OpenMP [4, 6]. Until recently, the programming model was a relatively simple flavor of the fork-join model: independent tasks were grouped in so-called parallel regions. All tasks within a region could be executed concurrently on OpenMP threads, while different regions were synchronized according to program order. In fact, the term task did not play a major role in the OpenMP specification. Starting with version 3.0 however, concepts such as explicit and untied task, data-dependency between asynchronous tasks, and execution target for offloading of tasks have successively enriched the OpenMP programming model.

Some debuggers, such as gdb, still perceive an OpenMP application as a collection of low-level threads, without any further semantics. Other debuggers, in particular those used in HPC such as DDT [1] and Totalview [13], have some awareness of the model and allow, for instance, to step all threads in a region as a group. TEMANEJO [5, 10] is a graphical debugger for task-based programming models such as OpenMP, OmpSs [7, 9, 11], StarPU [2, 3]. This tool follows a strongly model-centric approach and abstracts the application in terms of a task dependency graph, i.e. a directed graph consisting of tasks and data-dependencies as node and edges respectively. The user may inspect task inputs and outputs, or control the runtime system to step through the application task-wise or even modify task dependencies, thus using the same abstraction level and vocabulary as during program development.

This paper describes the Tasking Control API (TCA), a generic interface to interact with the runtime system of task-based programming models. The interface defines a small set of control requests, which need to be implemented by a conforming runtime system. These request allow a tool, for instance a debugger, to instruct the runtime system to suspend and resume task execution and to manipulate task-dependency. TCA has been designed to interoperate with the OpenMP Tools Interface (OMPT, [8]) and is currently used by the backend of the debugger TEMANEJO to interact with the OmpSs runtime system. The interface however, is sufficiently generic to be supported by other similar task-based programming models.

The paper is structured as follows: in Sect. 12.2 we summarize briefly the background for this paper, in Sect. 12.3 we present the design and specification of TCA, and Sect. 12.5 we draw some conclusions and give and outlook on future work.

12.2 Background

12.2.1 Temanejo

TEMANEJO is a graphical task based debugger. The tool consists of two parts, TEMANEJO (representing the frontend and the visible part for the user) and AYUDAME (the backend and the part of the tool chain talking with the runtime). For a given runtime we have to translate runtime specific events (creation of tasks, entering a parallel region, etc.) into AYUDAME events (Fig. 12.1) this is done inside the AYU-

Events — monitor runtime

- task creation/execution/shutdown
- dependencies/synchronization between tasks
- properties of tasks and dependencies (e.g. priority, owner, etc.)

Fig. 12.1 The runtime informs Ayudame about any relevant changes. The events are forwarded to Temanejo

Request — control runtime

- set/delete breakpoints
- execute/schedule next task (stepping)
- continue/break
- add/remove dependencies between tasks

Fig. 12.2 Temanejo controls the runtime/application via Ayudame by sending request. Requests are forwarded to the runtime

dame instrumentation library and mostly part of the runtime. The Ayudame specific events are transferred through socket communication towards Temanejo where they are displayed. For the purpose of debugging we also have to interact with the runtime. This is done by sending requests (Fig. 12.2) from Temanejo towards Ayudame through the socket communication. These requests now have to be translated into runtime specific behavior.

12.2.2 OMPT

For OmpSs and Ayudame the runtime-side implementation is done inside the Ayudame instrumentation plugin. As TCA is designed as an extension of OMPT and is at the moment implemented and used inside Ayudame but this code could later become part of the OMPT instrumentation plugin.

The initialization of OMPT is can be split into two stages. In the first stage the runtime calls an extern public initializer function this function is implemented inside the tool. As parameter the lookup function pointer is passed to the initialize function. Now the tool can ask for the different control request callbacks through the lookup function. As parameter for the lookup function a char pointer is passed. Depending on this string the lookup functions returns the requested inquiry function. This concept is illustrated in Fig. 12.3.

In the Fig. 12.4 we show the workflow of the second stage. In this stage the tool is instantiating callbacks for the different events in the runtime. This is done through the formerly requested inquiry function ompt_set_callback. The tool can now repeat this procedure to register all the events it wants to track.

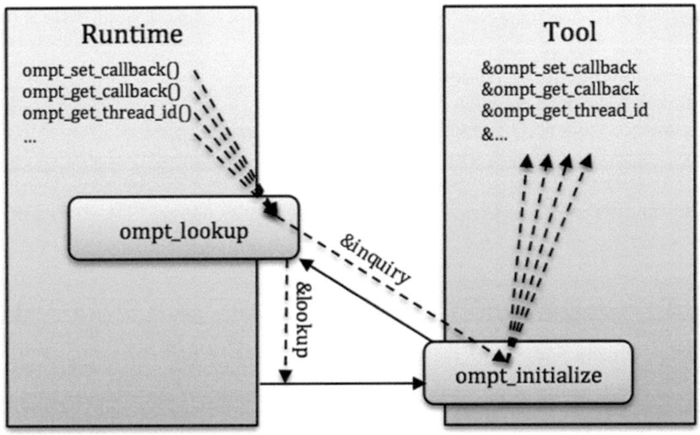

Fig. 12.3 Stage one of the OMPT initialization

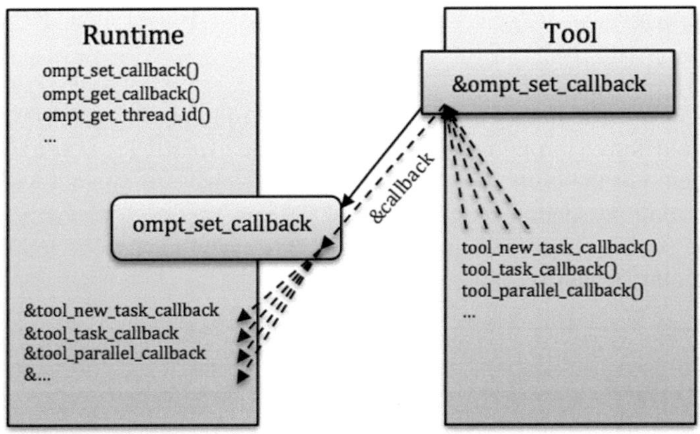

Fig. 12.4 Stage two of the OMPT initialization

12.3 Tasking Control Interface

In this chapter we describe the design of the Tasking Control Application Programming Interface (TCA).

12.3.1 Preliminaries

The purpose of TCA is to allow a tool to request certain actions from a conforming runtime system including:

- suspend and resume execution of a specific, labeled task
- insert and delete a dependency between two specific, labeled task

Some requests are mandatory, thus providing a minimal functionality to the tool, while other are optional. The foremost design objective is providing sufficient functionality to allow the debugger TEMANEJO to step through an application task-wise. A second major design objectives are interoperability with OMPT. In particular the design should meet two requirements:

- the procedure to initialize TCA should be the same as for OMPT
- do not introduce any concepts beyond those already present in OMPT, i.e. tool callbacks, runtime inquiry functions, tool data structures, etc.

Any runtime supporting OMPT should be able to support TCA trivially as long as it can provide the minimal set of TCA request. The design of TCA, however, is sufficiently generic to be applied with any runtime system for asynchronous task parallelization.

It is assumed, that the programming model meaningfully defines the concept of task and the concept of dependency. A program is composed of tasks instances and dependency instances between them, thus forming a task-dependency graph. A unique identifier can label task and dependency instances. Note, that the same dependency instance, and thus identifier, may appear more than once at different positions in the task-dependency-graph.[1] This might be the case when dependencies arise through data-dependencies; the same datum can be an input dependency for several tasks. In addition, we assume that the runtime system executes tasks on same kind of uniquely labeled execution resource. TCA uses the name thread for this concept; however, this does not imply that actual OS threads need to be used by the concrete runtime system. As with tasks and dependencies, threads have a unique identifier that can be used to identify it.

The purpose of TCA is to send control requests from a tool to a runtime system, only. It does not provide means to monitor the runtime system or inquire its state. This needs to be done through a separate channel—for instance through event callbacks in OMPT, or any other monitoring interface. In particular, the monitoring system needs to report task, dependency and thread identifiers, respectively.

Some aspects of the TCA are mandatory for conforming runtime systems or tools and thus form a minimal set allowing basic control. In addition TCA defines a range of optional features that allow more fine-grained and complex control. The specification of TCA is composed of:

- tool data structures
- control requests—functions implemented by the runtime system and invoked by the tool to request certain actions from the runtime system
- initialization—a prescription how the runtime system registers with the tool and exposes its capabilities

[1]The rest of this chapter will use task and dependency instead of task instance and dependency instance, respectively. Similarly, we refer to task-dependency graph as task-graph or graph.

12.3.2 Tool Data Structures

Task identifier **mandatory**
Each task is known to TCA by an identifier of type `tca_task_id_t`.

```
typedef uint64_t tca_task_id_t;
```

The task identifier is unique across all task instances and may appear only once in the dependency graph. The task identifier is used to name a specific task when requesting task related control actions from the runtime system. Tools have to learn the task identifier by means other than TCA, for instance a dedicated monitoring system or OMPT. The value `0` is reserved to indicate an invalid or unknown task.

Dependency identifier **mandatory**
Each dependency is known to TCA by an identifier of type `tca_dependency_id_t`.

```
typedef uint64_t tca_dependency_id_t;
```

The dependency identifier is unique across all dependency instances. However, dependency instance and thus dependency identifiers may appear more than once in the dependency-graph. The dependency identifier is used to name a specific dependency when requesting dependency related control actions from the runtime system. Tools have to learn the dependency identifier by means other than TCA. The value `0` is reserved to indicate an invalid or unknown dependency.

Thread identifier **mandatory** Each thread is known to TCA by an identifier of type `tca_thread_id_t`.

```
typedef uint64_t tca_thread_id_t;
```

The thread identifier is unique across all thread instances. The thread identifier is used to name a specific thread when requesting thread related control actions from the runtime system. Tools have to learn the thread identifier by means other than TCA. The value `0` is reserved to indicate an invalid or unknown thread.

Error code **mandatory** Each TCA control request function will return a value of type `tca_success_t` to indicate the outcome of the request.

```
typedef enum {
        TCA_SUCCESS,
        TCA_FAIL,
        TCA_UNKNOWN_REQUEST
} tca_success_t;
```

The value `TCA_UNKNOWN_REQUEST` is returned when the request is malformed or cannot be interpreted by the runtime system. For example inconsistent request arguments, or unknown identifiers. The value `TCA_SUCCESS` is returned if the request is well formed and can be satisfied by the runtime system. The value

TCA_FAIL is returned if the request is well formed but cannot be satisfied by the runtime system.

12.3.3 *Control Request Functions*

Control request functions are implemented by the runtime and called by the tools. These functions are not publicly visible. Instead, tools can acquire pointers to control request functions through the lookup mechanism during tool initialization as described in Sect. 12.3.4. Each control request function returns an error code as defined in the previous section. Some control requests are mandatory to ensure minimal tool functionality, others are optional.

Continue control request **mandatory**

```
tca_success_t tca_request_continue();
```

Instructs the runtime to continue normal execution of tasks until a breakpoint is reached or until a *break* control request is issued. This request has no arguments.

Break control request **mandatory**

```
tca_success_t tca_request_break();
```

Instructs the runtime system to stop executing any task until a *continue* or *step* control request is issued. This request has no arguments.

Step control request **mandatory**

```
tca_success_t tca_request_step();
```

Instructs the runtime system to execute a single task. The task should be chosen according to the runtime systems regular scheduling policies, even if a breakpoint is set on this particular task.

Block task control request **optional**

```
tca_success_t tca_request_block_task(
                  tca_task_id_t task_id);
```

Instruct the runtime system to block execution of the task with identifier task_id. The runtime system should ignore the named tasks for all scheduling decisions until an *unblock* request is issued for it.

Unblock task control request **optional**

```
tca_success_t tca_request_unblock_task(
                  tca_task_id_t task_id);
```

Instruct the runtime system to unblock the task with identifier `task_id`. The runtime should re-enable normal scheduling operation for the named task.

Run task control request **optional**

```
tca_success_t tca_request_run_task(
                tca_task_id_t task_id);
```

Instruct the runtime system to force execution of the task with identifier `task_id`. The runtime system should execute the task as soon as possible.

Insert dependency control request **optional**

```
tca_success_t tca_request_insert_dependency(
                tca_task_id_t from_task_id,
                tca_task_id_t to_task_id);
```

Instruct the runtime system to insert a dependency from task with identifier `from_task_id` to task with identifier `to_task_id`. The runtime system should not execute task with identifier `to_task_id` before task with identifier `from_task_id` has finished.

Remove dependency control request **optional**

```
tca_success_t tca_request_request_remove_dependency(
                tca_dependency_id_t dependency_id);
```

Instruct the runtime system to remove a dependency with the identifier `dependency_id`. The runtime system is now allowed to execute the `to_task` before the `from_task`, if there are no other dependencies.

Break at task control request **optional**

```
tca_success_t tca_request_break_at_task(
                tca_task_id_t task_id);
```

Instruct the runtime system to stop execution and scheduling of any other task when reached the execution of task with identifier `tca_task_id`. The runtime system is not allowed to execute or schedule any other task.

Unbreak at task control request **optional**

```
tca_success_t tca_request_unbreak_at_task(
                tca_task_id_t task_id);
```

Instruct the runtime system to remove the break at task flag from task with identifier `tca_task_id`. The runtime system is now allowed to schedule and execute tasks.

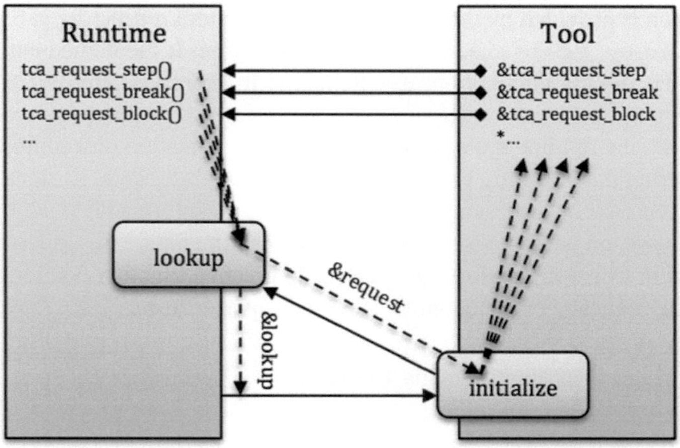

Fig. 12.5 Interaction between the runtime and the tool, though TCA

12.3.4 Initialization of Tools

Any tool using TCA needs to follow a specific initialization procedure as described in this section. The general principle (as illustrated in Fig. 12.5) is simple: the runtime calls an initialization function implemented by the tool. Within this function, the tool repeatedly calls a lookup function, which is provided by the runtime system, to inquire pointers to any control request function it wishes to use. If initialization fails, for instance because the runtime does not support a control request which is critical for the tool's operation, the tool returns with an error code from the initialization function.

The public interface of TCA is very slim; the only visible symbol is the initialisation function `tca_initialize`, which has the prototype definition

```
extern "C" {
tca_success_t tca_initialize(
                    tca_function_lookup_t lookup,
                    const char *runtime_version,
                    unsigned int tca_version);
}
```

All other symbols, i.e. control request functions, are publicized during TCA initialization through a lookup mechanism as described below. The second argument, `runtime_version`, is a version string that unambiguously identifies a runtime system implementation. The third argument, `tca_version`, indicates the version of the tasking control API supported by the runtime. The version of TCA described by this document is known as version 1. The first argument, `lookup`, is a pointer to a function with call signature

```
void * lookup(const char *entry_point);
```

This function is provided by the runtime system. The tool repeatedly calls `lookup` for every `entry_point`, i.e. control request function. If the named entry point is available, the lookup function returns a pointer to it. Otherwise, the `NULL` pointer is returned.

In general, the runtime system should start the initialization procedure as soon as possible during its on startup procedure and before any user code is executed. If the tool initialization is unsuccessful, i.e. if the tool does not return with `TCA_SUCCESS`, the runtime system need not maintain any information or state to support tools.

Use this in example section For example to obtain a function pointer, i.e. entry-point, to tca_request_continue, one invokes lookup as follows:

```
tca_interface_fn_t tca_request_continue_ptr =
                    lookup(tca_request_continue);
```

12.3.5 *Implementation By Example*

In this section we sketch out some aspect of the TCA implementation on the tool side as well as runtime system side. For illustration, we will use the fictional task-based programming model "MyRT" and the tool AYUDAME. For readability, we will ignore some of the necessary explicit cast operation, in particular on function pointers. The initialization function implemented by the tool could look something like

```
#include "tca.h"

tca_success_t tca_initialize(tca_function_lookup_t lookup,
      const char *rt_version, unsigned int tca_version) {


  if (TCA_VERSION != tca_version) {
    dprint("Incompatible TCA version\n");
    return TCA_FAIL;
  }

  if (0 == strncmp(rt_version, "MyRT v0.8")) {
    dprint("Found buggy runtime. Doing workaround\n");
    workaround_bug();
  }

  ayu_req_continue = lookup("tca_request_continue");
  ayu_req_break = lookup("tca_request_break");
  ayu_req_step = lookup("tca_request_step");
  ayu_req_run = lookup("tca_request_run_task");

  if ((NULL = = ayu_req_continue) || (NULL = = ayu_req_break) \
        || (NULL = = ayu_req_step)) {
    dprint("Minimal requirements are not met\n");
    return TCA_FAIL;
```

```
    if (NULL = = ayu_req_run) {
      dprint("Doing workaround for run_task request\n");
      workaround_run_task();
    }

    // all good
    return TCA_SUCCESS;
  }
```

First, the tool checks if the TCA version used by the runtime system and the tool, respectively, match. If it does not, it immediately aborts initialization, returning `TCA_FAIL`. Next, it checks for a known buggy runtime version and, if present, kicks-off some workaround to deal with it. The next step is to retrieve pointers to several control request functions and store them in variables that are visible to the rest of the tool. Some request functions are critical for tool operation, in this example it is just the mandatory set. If any of the critical control request is not implemented by the runtime system, the tool again aborts with a failure. Other control requests are not critical to tool operation, but require special action if not present. In the example it is `tca_request_run_task` that requires a special treatment.

The runtime system needs to implement a lookup function similar to the following

```
void * myrt_lookup(const char *entry_point) {
  if (0 == strncmp(entry_point, "tca_request_step")) {
    instrument_step();
    return &myrt_request_step;
  }

  // more control requests
  if (...) {...}

  // at this point the control request is unknown
   return NULL;
}
```

The lookup function checks the contents of `entry_point` against the list of available request functions. If it finds a match, the runtime might need to record, that the tool will use this control request, and then returns a pointer to the request function. If no match is found, the particular request function is not implemented and the lookup returns a `NULL` pointer.

The runtime system also has to make sure that the TCA initialization function is called during startup.

```
#include "tca.h"

void myrt_start() {
  // many other things to do

  // TCA init
  if (tca_initialize) {
    tca_init = tca_initialize(myrt_lookup, myrt_version, TCA_VERSION);
    if (TCA_SUCCESS != tca_init) {
      dprint("Failed to initialize TCA. Not instrumenting.\n");
```

```
            instrument_none();
        }
    }

    // other things
}
```

If initialization of the tool fails, the runtime system takes note of this fact in order to avoid unnecessary instrumentation, which might interfere with normal operation or lower performance. Finally, the runtime makes needs to make sure that scheduling of tasks, etc. honors the control requests issued by the tool. This could, for instance, be accomplished by implementing a finite state machine as part of the scheduler.

12.4 Debugging for OmpSs with TEMANEJO Through TCA

We have used the TCA as described in the paper to connect our debugger TEMANEJO, or more specifically its backend AYUDAME, with the runtime of the task-based parallel programming model OmpSs. Figure 12.6 shows the general structure of information flow. In our current prototype, the OmpSs runtime has been instrumented to emit plain AYUDAME events (Fig. 12.1). In the future we will use the OMPT interface of OmpSs to generate AYUDAME events. AYUDAME events are processed and ultimately forwarded to the frontend to visualize the task dependency graph. Part of the information transmitted in these events is identifiers for dependency and task instances, which are later used in TCA calls.

In the other direction, user interaction in TEMANEJO results in AYUDAME control request (Fig. 12.2). So far, these control requests have been used to drive a simple state-machine inside AYUDAME, which either suspends or resumes worker threads of the OmpSs runtime. Specifically, the state-machine determines which threads

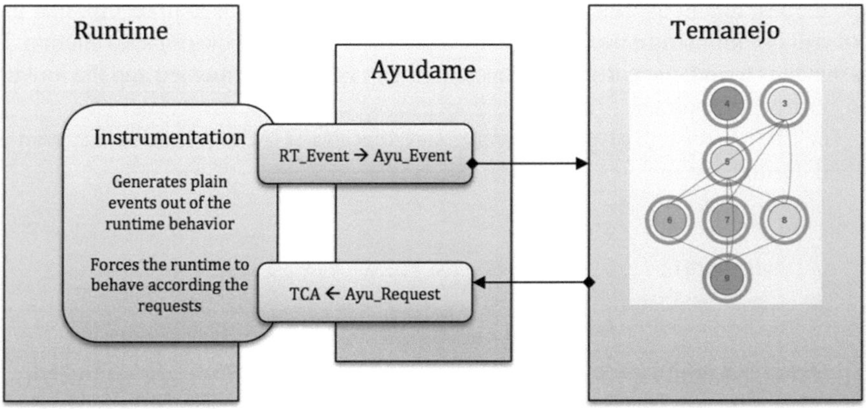

Fig. 12.6 Interaction of TEMANEJO with a runtime system through TCA

is allowed to return immediately from a AYUDAME event handler (and continue processing user tasks), or otherwise forced to busy loop inside AYUDAME(and thus break at certain tasks). The drawback of this approach is the assumption, that threads issuing AYUDAME events are those, which execute user tasks. While this assumption is true for OmpSs, it will not be in general.

Instead, the AYUDAME library will now generate TCA control request from AYUDAME ones. We have developed a plugin module for OmpSs, which initialises TCA as described in the previous section. The module also provides all mandatory and optional TCA control request functions and delegates serving all possible AYUDAME control actions to the OmpSs runtime system.

The services inside the OmpSs runtime do not allow to implement all TCA requests. For instance, there is no interface to prevent the scheduler from executing a specific task as required by `tca_request_block_task`. As a work around until such services are added to the OmpSs runtime, we have basically broken the simple state-machine described above out of the AYUDAME library and moved it into the OmpSs plugin. Our prototype is able to fully control the execution of an OmpSs application.

12.5 Conclusions

In this paper we present TCA, an interface that allows tools to control various aspects of the execution of task-based programming models as OpenMP or OmpSs. We use TCA in the backend of the graphical debugger TEMANEJO. This allows us for instance to step through a task-based application task-by-task and thus getting closer to the model-centric debugging approach. Internally, AYUDAME uses only TCA requests. If a runtime system does not support TCA, AYUDAME wraps specific runtime handlers inside TCA conforming control request.

TCA has been designed to interoperate with the OpenMP Tools Interface. Specifically, the initialization procedure of TCA is the same as OMPT. In fact, from the point of view of initialization, TCA's control request functions can be treated as OMPT's inquiry functions. Both need to be obtained by the tool through the same lookup mechanism. Any tool already supporting OMPT, could in addition lookup any TCA control request function. This is an obvious path for a future integration of TCA into OMPT. In the meantime, initialization of TCA and OMPT can be done separately. In AYUDAME, we have implemented TCA initialization function as a weak alias for the OMPT initialization function. During initialization, we lookup all any symbol, either TCA or OMPT, that is required for the operation of TEMANEJO.

We designed TCA in a generic way to interact with runtime systems for task-based programming models but is also sufficiently generic to be used with other programming models for asynchronous task parallelization. The design is easily extendable with new functionalities, to support the programmers needs.

References

1. Allinea: Allinea ddt: The debugger for c, c++ and f90 threaded and parallel code, http://www.allinea.com/
2. Augonnet, C., Thibault, S., Namyst, R., Wacrenier, P.A.: Starpu: a unified platform for task scheduling on heterogeneous multicore architectures. In: Proceedings of the 15th International Euro-Par Conference on Parallel Processing, Euro-Par '09, pp. 863–874. Springer, Berlin (2009)
3. Augonnet C, Thibault S, Namyst R, Wacrenier PA (2011) Starpu: a unified platform for task scheduling on heterogeneous multicore architectures. Concurr. Comput.: Pract. Exp. 23(2):187–198
4. Board, T.O.A.R.: The openmp api specification for parallel programming, http://openmp.org/
5. Brinkmann, S., Gracia, J., Niethammer, C.: Task debugging with temanejo. In: Cheptsov, A., Brinkmann, S., Gracia, J., Resch, M.M., Nagel, W.E. (eds.) Tools for High Performance Computing 2012, pp. 13–21. Springer, Berlin (2013)
6. Dagum L, Menon R (1998) Openmp: an industry standard api for shared-memory programming. IEEE Comput. Sci. Eng. 5(1):46–55
7. Duran A, Ayguadé E, Badia RM, Labarta J, Martinell L, Martorell X, Planas J (2011) Ompss: a proposal for programming heterogeneous multi-core architectures. Parallel Process. Lett. 21(02):173–193
8. Eichenberger, A., Mellor-Crummey, J., Schulz, M., Copty, N., Cownie, J., Dietrich, R., Liu, X., Loh, E., Lorenz, D.: Openmp technical report 2 on the ompt interface, http://openmp.org/mp-documents/ompt-tr2.pdf
9. Ferrer, R., Planas, J., Bellens, P., Duran, A., Gonzalez, M., Martorell, X., Badia, R.M., Ayguade, E., Labarta, J.: Optimizing the exploitation of multicore processors and gpus with openmp and opencl. In: Proceedings of the 23rd International Conference on Languages and Compilers for Parallel Computing, LCPC'10, pp. 215–229. Springer, Berlin (2011)
10. Keller, R., Brinkmann, S., Gracia, J., Niethammer, C.: Temanejo: Debugging of thread-based task-parallel programs in starss. In: Brunst, H., Mller, M.S., Nagel, W.E., Resch, M.M. (eds.) Tools for High Performance Computing 2011, pp. 131–137. Springer, Berlin (2012)
11. Perez, J., Badia, R., Labarta, J.: A dependency-aware task-based programming environment for multi-core architectures. In: Proceedings of the 2008 IEEE International Conference on Cluster Computing, pp. 142–151, September (2008)
12. Pouget, K., López Cueva, P., Santana, M., Méhaut, J.F.: Interactive debugging of dynamic dataflow embedded applications. In: Proceedings of the 18th International Workshop on High-Level Parallel Programming Models and Supportive Environments (HIPS). Boston, Massachusetts, USA, May (2013), held in conjunction of IPDPS
13. RogueWave: Faster fault isolation, improved memory optimization, and dynamic visualization for your high performance computing apps, http://www.roguewave.com/

Chapter 13
Evaluating Out-of-Order Engine Limitations Using Uop Flow Simulation

Vincent Palomares, David C. Wong, David J. Kuck and William Jalby

Abstract Out-of-order mechanisms in recent microarchitectures do a very good job at hiding latencies and improving performance. However, they come with limitations not easily modeled statically, and hard to quantify exactly even dynamically. This paper will present Uop Flow Simulation (UFS), a loop performance prediction technique accounting for such restrictions by combining static analysis and cycle-driven simulation. UFS simulates the behavior of the execution pipeline when executing a loop. It handles instruction latencies, dependencies, out-of-order resource consumption and other low-level details while completely ignoring semantics. We will use a UFS prototype to validate our approach on Sandy Bridge using loops from real-world HPC applications, showing it is both accurate and very fast (reaching simulation speeds of hundreds of thousands of cycles per second).

13.1 Introduction

Different performance models can be built with very different purposes in mind. For instance, CPU architects may need very low-level cycle-accurate simulators to find and fix bugs in their design, in which case accuracy is so important that practical aspects like processing time and lightweightness become completely secondary. At the other end of the spectrum, models like Intel Architecture Code Analyzer (IACA) [1], Code Quality Analyzer (CQA) [2] and Cape [3] aim to provide good-enough predictions at minimum cost, both in terms of time and space. Many models have been developed as compromises between these two extremes.

In this paper, we will present *Uop Flow Simulation* [4] (UFS), a technique to accurately evaluate a loop's performance for L1 data sets by simulating the behavior of a CPU's out-of-order engine on a cycle-accurate basis, and offering the following advantages:

V. Palomares (✉) · W. Jalby
University of Versailles Saint-Quentin-en-Yvelines, Versailles, France
e-mail: vincent.palomares@uvsq.fr

D.C. Wong · D.J. Kuck
Intel Corporation, Champaign, IL, USA

A. Knüpfer et al. (eds.), *Tools for High Performance Computing 2015*,
DOI 10.1007/978-3-319-39589-0_13

1. Modeling key components of the out-of-order engine (issue, buffers, dispatch...) with a reasonable accuracy.
2. Very fast speed: several hundred thousand simulated cycles per second in typical cases.
3. Low memory consumption: only a few MB of RAM are required.
4. Small input files: only limited, statically extracted information needs to be used for a given loop.

We will present a motivating example to showcase the value of UFS and describe our model in details. Then, we will validate our implementation on real-world loops from the AVBP [5] and YALES2 [6] industrial applications, targeting the Sandy Bridge microarchitecture.

We will use CQA as a reference point to highlight our model's contribution. CQA leverages a performance model close to the one developed in IACA, operating in a mostly bandwidth-centric fashion and overlooking the impact of out-of-order buffers. Both CQA and our UFS prototype operate at the binary loop level and rely on static information extracted using the MAQAO [7] framework to perform their analyses, making them easy to compare.

13.2 Motivating Example

Realft2_4_de (Fig. 13.1a) is a codelet from our Numerical Recipes suite, and is part of an inverse Fourier transform algorithm. It is a particularly interesting codelet as it suffers from an important CQA error even for L1 data sets. Indeed, CQA overestimates

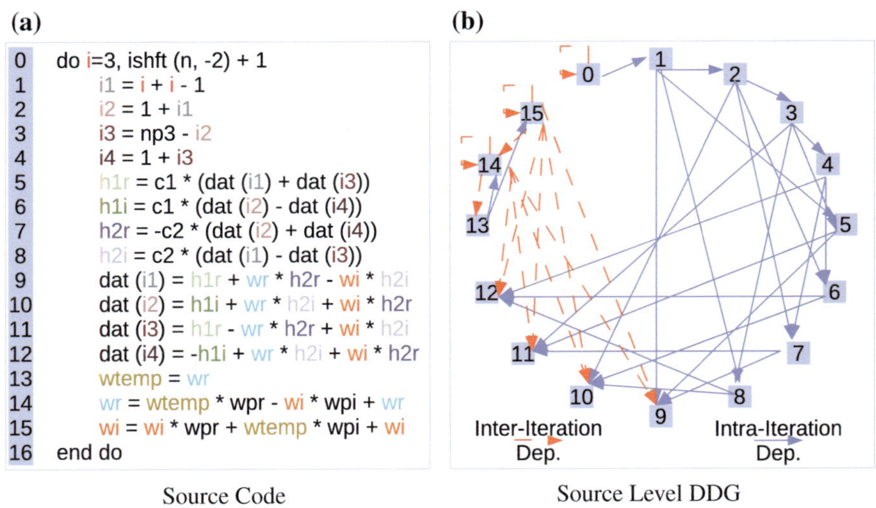

Fig. 13.1 Realft2_4_de Codelet—Realft2_4_de is a codelet from our Numerical Recipes suite, and is part of an inverse Fourier transform algorithm. The DDG (Data Dependency Graph) was made at source level for clarity purposes

Table 13.1 Realft2_4_de Measurements, CQA and UFS Projections—The target microarchitecture is Sandy Bridge

Source	Cycles	Stalls	Error (Cycles)
Measured	23.36	7.85 [RS]	N/A
CQA	16.00	N/A	31.51 %
UFS (Normal buffers)	23.01	19.97 [RS]	1.50 %
UFS (Large buffers)	19.03	5.65 [ROB]	N/A

Durations and stalls represents values per assembly loop iteration. The *error* metric is defined as: $error = |\frac{measured\ execution\ time - predicted\ execution\ time}{measured\ execution\ time}|$. While the simulated stalls count the number of cycles where at least 1 uop could not be issued due to a buffer being full, what hardware counters exactly measure is not very clear (and could e.g. be the number of cycles where no uop at all was issued), explaining the importance of the gap between measured and simulated stall counts

the loop's speed by roughly 45 % (see Table 13.1), making it a good case for detailed study.

Detailed investigation using hardware counters revealed that stalls due to out-of-order resource scarcity may have had an impact on the actual performance. While not all stalls necessarily impede the execution, this was still an interesting lead directly incriminating the Reservation Station (RS): as it holds uops until their operands are ready, it gets particularly stressed when there are dependencies between instructions. We gave them a particular scrutiny in Fig. 13.1b.

Running UFS on the target loop gives encouraging results (see Table 13.1), completely filling the gap between the CQA cycles estimation and the measurement.

We can notice a 31 % gap between the measured time per iteration and the bandwidth-centric CQA projection. Measured resource stall counters indicate the Reservation Station (not modeled by CQA) may have been hindering performance, though they do not guarantee it is the case.

The error for UFS (with regular Sandy Bridge parameters) is of only 1.50 %, a significant improvement over CQA's 31.51 %. This accuracy gain is due to two main factors:

1. RS size awareness: running UFS again with a virtually infinite RS size shows that 4 cycles could be gained from having larger out-of-order buffers (Table 13.1).
2. Realistic dispatch: hardware heuristics prioritized uops that are not on the critical path.

Furthermore, running UFS again with all out-of-order resources set to have 1000 entries (*UFS - Large Buffers* row) shows nearly 4 ($23.01 - 19.03 = 3.98$) cycles can be gained by merely increasing buffer sizes, demonstrating that buffers do indeed limit performance here.

The CQA evaluation represents the performance the loop would attain if not for these issues.

13.3 Model Presentation

The purpose of the model is to account for limitations of the out-of-order engine not taken into account in CQA using a limited cycle-accurate simulation. The semantics of instructions is completely disregarded; only the flow of uops is being computed, estimating the speed at which uops may travel through the pipeline.

Detailed interactions between uops and the pipeline are taken into account for this purpose. For instance, the simulation keeps track of in-flight uops and the number of available resources, constraining the simulated flow of uops as a real system would. Dispatching constraints and heuristics are also implemented, allowing for a realistic estimation of port load in complex loops.

13.3.1 Limited Input

The simulator uses two types of inputs:

1. Loop information: as the model is only tracking the flow of uops and not their semantic purpose, register and memory values are not needed. It uses only basic information obtainable from static analysis, such as the type, register operands and outputs for each instruction in the studied loop. It also uses Agner's instruction tables [8] as reference for instruction dispatch port(s) and latency. Just as with CQA, loop inputs are generated using the MAQAO [9] framework. An example input is provided in Table 13.2.
2. Microarchitecture information: simple parameters such as the size of each out-of-order resource or the Front-End and issue uop bandwidth are needed. Default values (see Table 13.3) can be provided for each target microarchitecture. All the studied microarchitectures have an issue and retire bandwidth of 4 uops per cycle. A few behaviors are also microarchitecture specific, such as the status of microfused uops in the ROB.

The number of loop iterations to simulate can also be specified, with a default value of 1000.

13.3.2 Model Overview

UFS simulates as many cycles as needed for *the last uop of the Nth iteration to retire*, with N being the number of iterations to simulate. The different simulation steps for a cycle are as follows:

1. *Issue*: inserts new uops in the ROB (and RS if need be) and allocates all needed resources. This step is done in order.

Table 13.2 Partial UFS Loop Input Example (Realft2_4_de)—The specifics of what an instruction exactly does is irrelevant for UFS

#Insn	Nb_FE	Type	Input	Output	Latency	Ports
1	1	compute	XMM1	XMM10	1	P5
2	1	compute	RAX	RDI	1	P1, P5
…						
8	1	load	RSI, RDI	XMM15	4	P2, P3
…						
53	1	store_addr	RSI, R8		1	P2, P3
		store	XMM15		3	P4
54	1	branch	RDX, RAX	test	1	P5

Instead, only characteristics such as the number of uops it takes in the Front-End (*Nb_FE*), input/output registers and latency are important. The *type* category allows us to determine what type of resource the uop will need to get issued (e.g. branch buffer entry for *branch* uops), coupled with the *output* field: for instance, instruction 1 is going to need a vector register as it produces an XMM value, while instruction 2 will need an integer one. The *ports* column provides a list of ports compatible with the uop. In some cases, a single instruction gets split into several uops. As each of them may potentially have different attributes, differences between them need to be described explicitly. This is the case for instruction 53. Complementary attributes are sometimes needed, e.g. in case of division uops, as they are going to use special resources exclusively for variable amounts of time

Table 13.3 Default Microarchitectural Input Parameters—We found the practical sizes of the PRFs, the ROB and the RS to be different from the official ones in all three microarchitectures [10]

Microarchitecture	Sandy Bridge	Ivy Bridge	Haswell
Front-End (FE) Bandwidth	4	4	4
Branch Buffer (BB) Entries	48	48	48
Load Buffer (LB) Entries	64	64	72
Load Matrix (LM) Entries	32	32	32
FP Physical Registers (FP - PRF)	112	113	138
Integer Physical Registers (Int - PRF)	128	130	144
Max Number of Allocated Registers (All - PRF)	141	165	177
ReOrder Buffer (ROB) Entries	165	168	192
Reservation Station (RS) Entries	48	51	51
Store Buffer (SB) Entries	36	36	42
ReOrder Buffer Microfusion	No	Yes	Yes
Reservation Station Microfusion	No	No	No

Such differences can be explained by e.g. the mode the processor is working in (64-bit mode exposes more named registers than the 32-bit one), resources needed to maintain the architectural state (e.g. 16 physical registers for the 16 named registers), the number of pipeline stages involved in allocating or releasing resources, or by technical limitations unknown to us. In any case, the simulator inputs should be the number of resources available for speculation

2. *Dispatch*: removes uops from the RS when all the uops they depend on were properly *executed* and a compatible execution port is available. This is done out of order.
3. *Update*: flags uops as being *executed* L cycles after they were dispatched, with L being their attributed latency. Also releases the Load Matrix entries used by executed load uops. This is also done out of order.
4. *Retire*: removes *executed* uops from the pipeline and releases the resources they still held. In a real microarchitecture, this is the step at which *executed* uops' outputs would be committed to the architectural state. Retirement is done in order.

The *current cycle count* (i.e. number of cycles simulated so far) is maintained and updated every cycle.

Figure 13.2 presents an overview of the model. We will describe details for each simulated component in the coming sections.

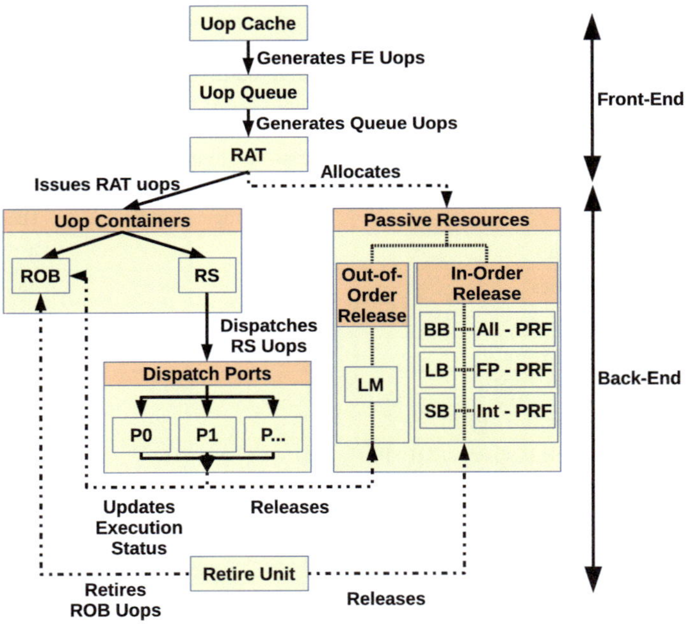

Fig. 13.2 UFS Uop Flow Chart—Several types of uops are used in the pipeline. For instance, Front-End uops (FE uops) are different from Queue uops or ROB uops. Generally, the closer to the Front-End, the closer to the original instruction the uop is. As different components can split uops when processing them, Back-End uops may contain less of the original information and semantic than their earlier counterparts. In other words, more Back-End uops than FE uops may be needed to describe the same instruction. Such transformations will be detailed in the matching component's modeling description. The ROB and the RS are both resources and uop containers. Other resources have a more passive role and do not describe uops, acting instead as mere dependencies

13.3.3 Simplified Front-End

As UFS targets loops, we can safely assume that all the uops sent to the *uop queue* come from the *uop cache*, hence ignoring the legacy decode pipeline and its limitations and providing a constant uop bandwidth of 4 per cycle. While the uop cache has limitations of its own (e.g. it cannot generate more than 32B worth of uops in a cycle), we decided to ignore them as we could not find real world cases where they got in the picture. This is partly due to compilers being smart enough to avoid dangerous situations by using code padding.

We also assume the branch predictor is perfect and never makes mistakes, meaning we do not need to simulate any roll back mechanisms. This is a decently safe assumption for the loops we study due to their high numbers of iterations, but reduces the applicability of UFS for loops with unpredictable branch patterns.

We consequently model Front-End performance in a simplified way:

1. For Sandy Bridge (SNB) / Ivy Bridge (IVB): 4 uops can be generated every cycle, except a uop queue limitation prevents uops from different iterations from being sent to the Resource Allocation Table (RAT) in the same cycle. For instance, if a loop body contains 10 uops, the uop queue will send 4 uops in the first two cycles, but only 2 in the third.
2. For Haswell (HSW): 4 uops can be generated every cycle: the limit experienced in SNB and IVB was apparently lifted.

In some cases, the uop queue has to unfuse microfused uops (or *unlaminate* uops) before being able to send them to the RAT [11], causing more issue bandwidth to be consumed (and sometimes, also more out-of-order resources). We take common cases into account using the following rules:

1. For SNB / IVB, unlaminate when the number of register inputs for the whole instruction is greater than 2.
2. For HSW, unlaminate when the number of register inputs and outputs *for an AVX instruction* is greater than 3. This rule was obtained empirically.

13.3.4 Resource Allocation Table (RAT)

The simulated RAT is in charge of *issuing* uops from the Uop Queue to the ROB and the RS, as well as allocating the resources necessary to their proper execution and binding them to specific ports.

It does not have any bandwidth limit other than the one induced by the Uop Queue's output.

Resource Allocation

In regular cases, resource allocation is quite straight-forward. For instance, all uops need a spot in the ReOrder Buffer (ROB), loads need Load Buffer (LB) and Load

Matrix (LM) entries, etc. However, it gets more complex when an instruction is decomposed into more than a single uop. In our implementation, all resources needed at the instruction level will be allocated when the first uop reaches the Back-End. For instance, stores are decomposed into a *store address* and a *store data* uops: in this case, a Store Buffer (SB) entry will be reserved as soon as the *store address* uop is issued, and the second uop will be assumed to use the same entry. However, individual uop resources (ROB or RS entry) will still be allocated at the uop granularity.

It is important to note that if any resource is missing for the uop being currently considered, the RAT will stall and not issue any other uop until resources for the first one are first made available. This is commonly referred to as a *resource stall*.

Port Binding

Available information about dispatch algorithms in recent Intel microprocessors is rare and limited. We decided to bind uops to single ports in the RAT, sparing the RS from having to do a complex cycle-per-cycle evaluation of dispatch opportunities. Smarter strategies could be used, but we preferred to keep our simulation rules as simple as reasonably possible.

The simulated RAT keeps track of the number of in-flight uops assigned to each port, and assigns any queue uop with several port options to the least loaded one. In case of equality, the port with the lowest digit is assigned (this creates a slight bias towards low-digit ports).

This process is repeated on a per-uop basis, i.e. the simulated RAT uses knowledge generated by issuing younger uops in the same cycle, rather than using counts only updated once a cycle, which may in turn be optimistic.

Arbitrary numbers of ports can be activated as their use is regulated by the loop input file anyway (see Table 13.2). New microarchitectures with more (or fewer) ports could be simulated by tweaking input files' uop port attribution scheme to match the target's.

13.3.5 Out-of-Order Flow

Reservation Station (Uop Scheduler)

When arriving in the Reservation Station, queue uops that are still microfused get split in two, simplifying the dispatch mechanism.

The RS holds uops until (a) their operands are ready, (b) the needed port is free and (c) the needed functional unit is available. When all conditions are met, the RS *dispatches* uops to their assigned port prioritizing older uops, releasing the RS entries they used.

Port and Functional Unit Modeling

Ports act as gateways to the functional units they manage. They are modeled as all being completely identical, and being able to process any uop sent to them by the

RS. Functional units are not modeled distinctly, and constraints over them are modeled inside their respective port instead. Several rules are applied to match realistic settings:

1. A port can only process a single uop per cycle (enforced by dispatch algorithm).
2. Uops can be flagged as needing exclusive use of certain functional units for several cycles. For instance, division uops will make exclusive use of the divider unit for (potentially) dozens of cycles. A port processing such a uop will flag itself as not being able to handle other uops needing this particular unit for the specified duration. The same mechanism is also used for 256-bit memory operations on SNB and IVB. A port with busy functional units can still service uops not needing them.
3. While the port itself does not check whether it should legally be able to process a given uop, the RS verifies this a priori, preventing such situations in the first place.

Uop Execution Status Modeling

ROB uops have a time stamp field used to mark their status, and holding the cycle count at which they will be fully executed. By convention, the default value for newly issued uops is -1: its output is available if $current\ cycle\ count \geq the\ uop's\ execution\ time\ stamp > -1$.

Updating ROB uops' execution time stamp is typically done at dispatch time: as we deal with constant latencies, we can know in advance on what cycle the uop's output is going to be ready ($current\ cycle\ count + uop\ latency$).

In the case of typical *nop*-typed instructions (such as *NOP* and zero-idiom instructions like *XOR %some_reg, %same_reg*), the time stamp is directly populated with a correct value at issue time, reflecting the RAT being able to process them completely in our target microarchitectures. As they also have 0 cycle of latency, their stamp is simply set to $current\ cycle\ count$.

However, we found an extra simulation step to be necessary to handle *zero-latency register moves* (implemented in IVB and HSW), which are *nop*-typed and are entirely handled at issue time too. Contrary to *NOP*s or zero-idioms, register moves have register inputs, the availability of which is not necessarily established yet when the move uop is issued. We tackle this issue by inserting such uops with a negative time stamp *if* their input operand's availability is not known yet, and letting a new "uop status update" simulation step update them when it is.

This new *update* step is also in charge of releasing Load Matrix entries allocated to the load uops whose execution was just completed.

13.3.6 Retirement

The retirement unit removes uops from the ROB and releases their resources (other than RS and LM entries, which were already freed earlier in the pipeline).

1. Retirement is done in-order: no uop can be retired if an older uop still exists in the ROB. This is necessary to be able to handle precise exceptions and rollback to a legal state.
2. The default retirement bandwidth is the same as the FE's (4 uops per cycle) to prevent retirement from being the bottleneck in terms of throughput. To ensure this, ROB uops that are still microfused in the RAT are only counted as a single uop for retirement purposes.
3. Resources released in a given cycle cannot be reused in the same cycle. Our understanding is that it would be extremely complex to implement a solution allowing this, with very little performance to be gained (potentially increasing each resource's effective size by a maximum of 4). Note: we apply the same reasoning to the RS and the LM, even though their entries are freed at dispatch time (for the RS) or update/completion time (for the LM) instead of at retirement.
4. Resources allocated at the instruction level at issue time are released when retiring the last uop for this specific instruction. This is consistent with the resource allocation scheme we use at the issue step.

13.3.7 Overlooked Issues

Many aspects of the target microarchitectures are not simulated. Some of them are inherently so due to our approach and the lack of dynamic information, such as cache and RAM behavior, Read after Write (RAW) memory dependencies and branch mispredictions.

Others are implementation choices and may be subject to change, like the number of pipeline stages (which could have an impact on the resource allocation scheme), the impact of yet-to-be-executed store address uops on later load and store uops, writeback bus conflicts [12] or partial register stalls [13].

Furthermore, while a lot of information is available concerning the way Intel CPUs work, many hardware implementation details are not publicly available. We could fill some of the gaps using reasonable guesses, but they are probably flawed to some degree, restricting the accuracy attainable by our model.

13.4 Validation

The validation work for UFS is twofold:

1. Accuracy: checking whether the model provides faithful time estimations for loops operating in L1.
2. Speed: making sure simulations are not prohibitively slow for their intended use.

We will focus on Sandy Bridge validation, as identifying performance drops on this microarchitecture was the primary motivation for developing UFS in the first place. Furthermore, its modeling is used as basis for IVB and HSW support, making SNB validation particularly important.

We will use the *fidelity* metric (defined here as $fidelity = 1 - error$) to represent UFS accuracy for each studied loop, and systematically compare UFS results with CQA projections to highlight our model's contributions.

A short study of the time taken by our UFS prototype will be made, and results will be presented in terms of *simulated cycles per second*.

13.4.1 Fidelity

Experimental Setup

The host machine had a two-socket *E5-2670* SNB CPU, with 32 KB of data L1 cache, 256 KB of L2, and 20 MB of L3. It also had 32 GB of DDR3 RAM.

For each tested application, we selected loops that:

1. Are hot spots: the studied loops are relevant to the application's performance.
2. Are innermost, have no conditional code and can therefore be analyzed out of context.
3. Have a measured time greater than 500 cycles per loop call. This is needed to make sure measurements are reliable (small ones can be inconsistent [14]). This may exclude small loops that are called numerous times.

Performance measurements were performed in vivo using the DECAN [15] differential analysis tool.

We use DECAN variant *DL1* to force all memory accesses to hit constant locations, and thereby getting a precise idea of what the original loop's performance would be if its working set fit in L1. This also allows us to make direct comparisons between measured cycles per iteration vs. UFS and CQA projections, as other components of the memory hierarchy are artificially withdrawn from the picture.

AVBP

AVBP [5] is a parallel CFD numerical simulator targeting reactive unsteady flows. Its performance scales nearly linearly for up to 4 K nodes.

Figure 13.3 shows UFS and CQA results for 29 AVBP hot loops on Sandy Bridge. UFS shows fidelity gains of more than 5 percentage points for 9 of them, with a maximum gain of 27 percentage points for loops 7507 and 7510. Other important gains include 20 percentage points for loops 7719 and 3665.

The worst fidelity for UFS is 78.18 % for loop 13906 (against 66.76 % for CQA on loop 3665).

The average fidelity is of 91.73 % for UFS, versus 86.34 % for CQA.

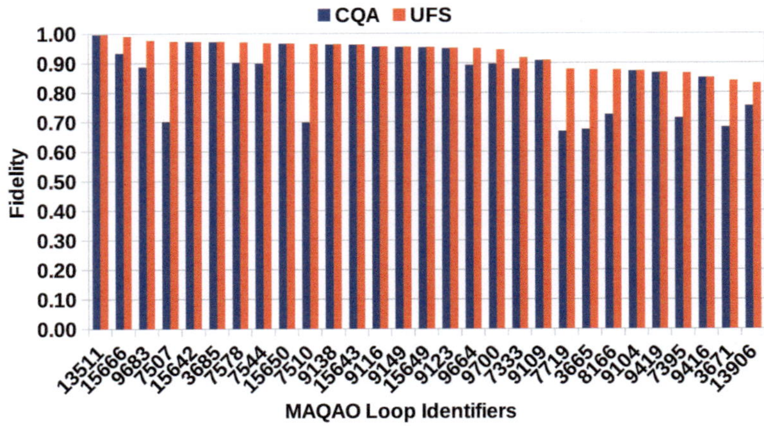

Fig. 13.3 In Vivo Validation for DL1: AVBP. Results are sorted by descending UFS fidelity

YALES2: 3D Cylinder

YALES2 [6, 16] is a numerical simulator of turbulent reactive flows using the Large Eddy Simulation method. Its performance scales almost linearly with the number of execution cores even with thousands of cores.

Figure 13.4 shows UFS and CQA results for the *3D cylinder* part of this application. UFS shows fidelity gains of more than 5 % points for 12 loops out of 26, with a maximum gain of 35 for loop 22062. Other particularly important gains include 28 and 24 % points for respectively loops 22040 and 4389.

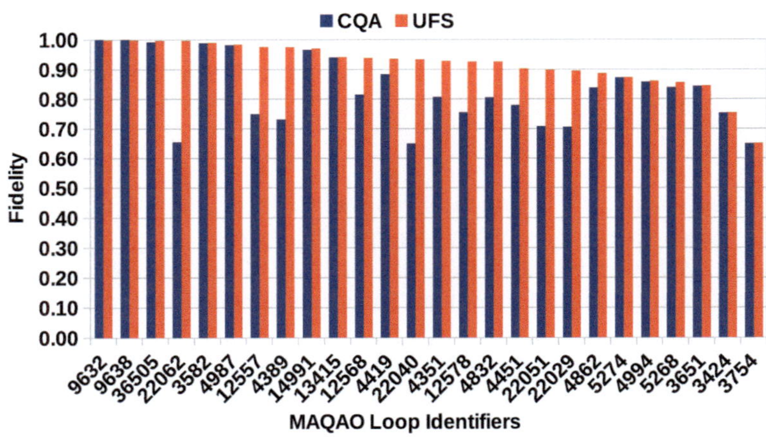

Fig. 13.4 In Vivo Validation for DL1: YALES2 (3D Cylinder). Results are sorted by descending UFS fidelity

Some loops' performance are impacted by factors apparently not modeled by UFS, with disappointing fidelities of respectively 65.01 % and 75.32 % for loops 3754 and 3424.

The average fidelity is of 91.67 % for UFS, versus 82.93 % for CQA.

13.4.2 Simulation Speed

Speed is very important for performance evaluation tools, especially in the context of optimization: various versions of a program can be tested, e.g. trying different compiler flags or hand optimizations.

The quality of a model can be thought of in terms of return on investment: are the model's insights worth their cost?

We will hence study UFS's speed in this section, and evaluate the cost of UFS analyses.

Experimental Setup

Simulations were run serially on a desktop machine with an *i7-4770* HSW CPU, running at 3.4 GHz. They were run on a single core, with 32 KB of L1 data cache, 256 KB of L2 cache and 8 MB of L3. It also had 16 GB of DDR3 RAM.

The targeted microarchitecture was SNB, with its default microarchitectural parameters, but simulating different numbers of iterations: 1000 and 100 000. The former is the default one and the most relevant to our analysis, while the latter was run to give an idea of sustained simulation speeds past the initialization phase (slowed down by I/O).

Execution times were measured using the *time* Linux tool, with a resolution time of 10 ms. While other measurement methods would be more precise, we deemed this one to be enough for our intended purposes. Furthermore, the time needed to generate the loop input files with MAQAO is not counted here. Measures were performed with 11 meta-repetitions to stabilize results.

AVBP

Figure 13.5 shows simulation speeds for the AVBP loops we studied. Here, the time needed to simulate 1000 iterations has a high variability, and can go from as low as .02 s for loop 3685 to as high as 2.73 s for loop 7578. The average simulation time is of around .28 s for each loop. This is due to the high complexity of some of the loops, which comprise hundreds of instructions (200 assembly statements on average). In the case of loop 7578, there are 1337 instructions (including divisions), making each of the 1000 iterations require many simulated cycles to complete. Hence, each iteration needs more simulated cycles to complete. Furthermore, the number of instructions can impact the locality of our UFS prototype's data structures, with large loops consequently being simulated less quickly.

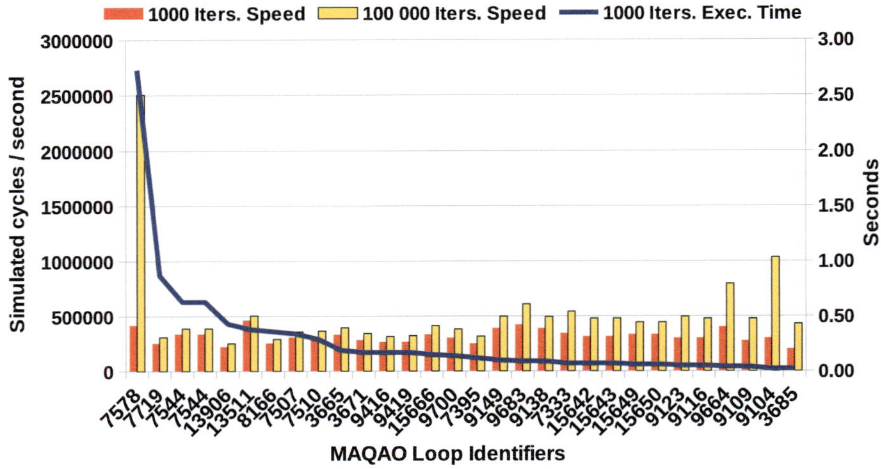

Fig. 13.5 UFS Speed Validation for AVBP. Results are sorted by descending UFS execution time. Our UFS prototype simulates an average of 300 K cycles per second for the studied loops. This average is 1.6x higher when simulating 100 000 iterations

For AVBP, we achieve on average:

1. Simulation times (for 1000 iterations) of approximately 0.28 s per loop: we can sequentially simulate around 3.57 loops per second.
2. The simulation of 318 K cycles per second for 1000 iterations (and 519 K for 100 000 iterations).

YALES2: 3D Cylinder

Figure 13.6 shows simulation speeds for the YALES2 (3D Cylinder) loops we studied.

As with AVBP loops, the simulation time for 1000 iterations is highly variable, going from 0.02 to 0.56 s. Simulations take 0.13 s on average, which is shorter than for AVBP (0.28 s). We can hence sequentially simulate an average number of $\sim$ 7.69 YALES2 loops per second. The difference is due to YALES2 loops being relatively less complex, with an average size of 110 assembly statements (against 200 for AVBP).

However, the average number of simulated cycles per second is similar, reaching 281 K cycles per second when simulating 1000 iterations –against 318 K cycles per second on AVBP– (respectively 549 K and 519K when simulating 100 000 iterations).

Comparison with CQA

We will quickly assess CQA's speed to compare it to UFS's. To do so, we ran CQA on the AVBP binary for all the loops studied earlier (in a single run).

When removing the overhead due to the MAQAO framework (mostly consisting in disassembling the binary) to make fair comparisons with UFS, we found that CQA could process 30.98 loops per second. We can hence roughly estimate UFS to be $30.98/3.57 \simeq 8.68$x for AVBP's hot loops.

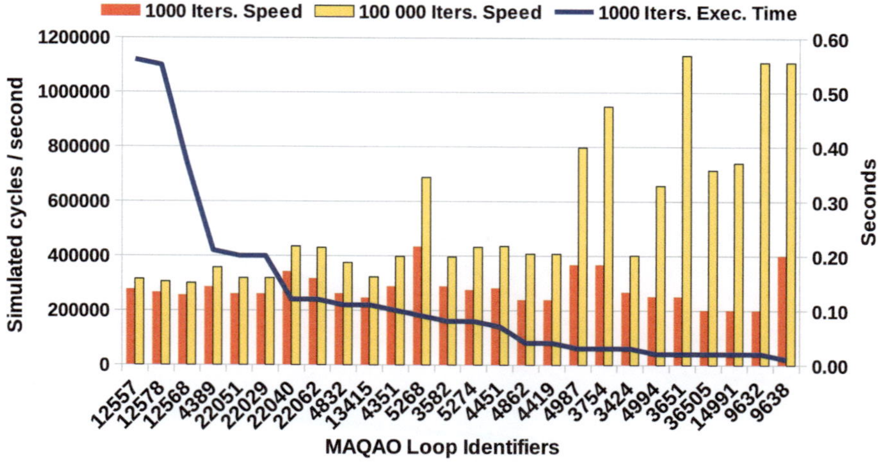

Fig. 13.6 UFS Speed Validation for YALES2: 3D Cylinder. Results are sorted by descending UFS execution time. Our UFS prototype typically simulates around 200 K cycles per second here. This number doubles when simulating 100 000 iterations. In practice, simulation times are around 0.10 s for each loop

Applied to YALES2, the same methodology shows that CQA can process 42.85 loops per second when targeting the hot loops we studied earlier. This brings the overhead for using UFS to $42.85/7.69 \simeq 5.57$x for YALES2's loop hotspots.

This difference is larger ($\sim$13x) for smaller loops, of which CQA can process around 280 per second, compared to approximately 21 with UFS (the detailed data is not presented in this paper).

Overall, UFS analyses take around 10x more time than CQA's.

13.5 Sensitivity Analyses

We can use UFS to perform sensitivity analyses and evaluate how loops of interest would behave given different microarchitectural inputs.

13.5.1 Latency Sensitivity Analysis

We will evaluate the behavior of different loops as the performance of the cache hierarchy varies. While our model does not support detailed cache modeling, we can still change L1 performance by changing the latency of load and/or store uops.

Figure 13.7 shows how different loops react to latency variation. We can notice a wide range of behaviors, with interesting outliers:

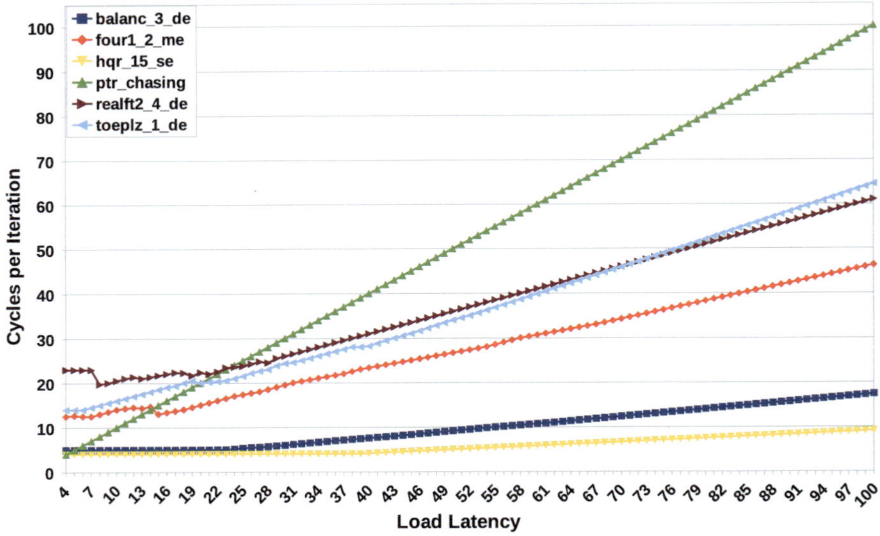

Fig. 13.7 Sensitivity Analysis: Load Latency. The unit of the Load Latency axis is cycles. The presented loops were extracted from the Numerical Recipes [3, 17], except for *ptr_chasing*. We can see loops can react very differently to latency increases, with pointer chasing being most impacted

- *ptr_chasing*: a loop chasing dependent pointers (1 pointer per iteration), and where only one load uop can consequently be executed in parallel. Load latency entirely governs its performance: its *Cycles per Iteration* metric scales perfectly with the latency of loads on the studied range of latencies (slope = 1x). This represents the worst case scenario for latency scaling.
- *hqr_15_se*: a loop with very few arithmetic operations per load, allowing it to execute many load uops in parallel. It can absorb important amounts of latency without getting degraded performance (up to 39 cycles), and then its *Cycles per Iteration* value scales only weakly with latency (slope $\simeq 0.09$x).

Furthermore, some loops surprisingly get better performance for higher latencies (which is most noticeable on *realft2_4_de*, where the number of *Cycles per Iteration* drops on point 8): the change in latency causes uops from the Reservation Station to become ready at different times, changing the order in which they get dispatched (and coincidentally reaching a better dispatch scheme than with a lower load latency). This only happens locally, though, and the regular behavior (of performance dropping as latency increases) gets back in the picture on later data points.

13.5.2 Resource Size Sensitivity Analysis

We can also easily quantify how sensitive a loop is to the size of out-of-order buffers, i.e. see the impact of buffer sizes on instruction level parallelism (ILP) for the studied loop.

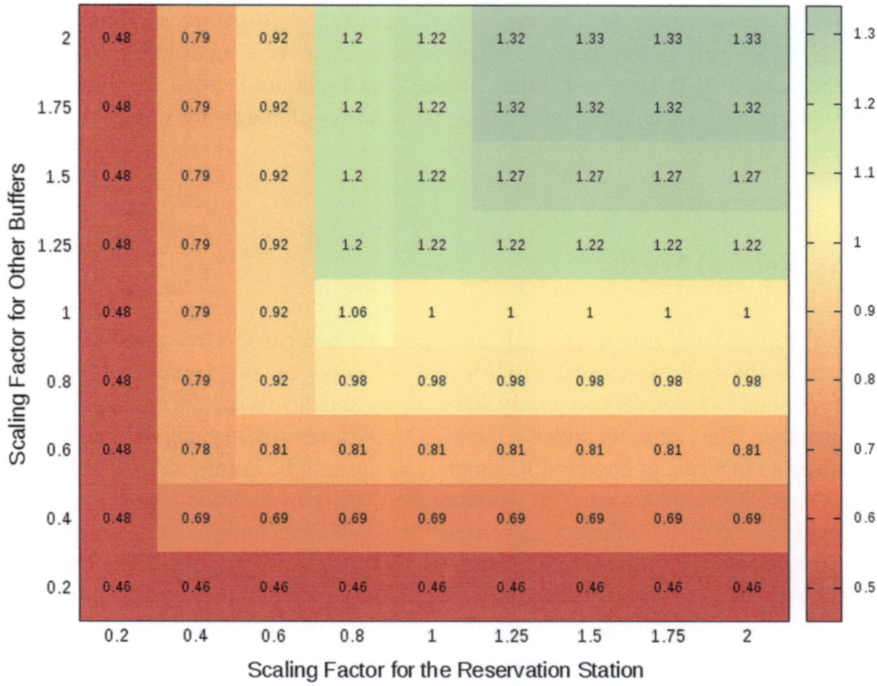

Fig. 13.8 Sensitivity Analysis: Resource Scaling Speedup (YALES2: Loop 4389). This heatmap represents the speedup obtainable when scaling the size of the Reservation Station or/and other out-of-order buffers, with regular Sandy Bridge parameters being used as reference (on coordinates (1, 1))

In Fig. 13.8, we evaluate how a loop's performance varies depending on the sizes of the RS and other buffers (and particularly the ROB). We can see that no speedup can be achieved from merely increasing the size of the RS (coordinates (2, 1)). However, increasing the size of other buffers (and particularly that of the ROB, in this case) by 25 % can provide a speedup of 1.22x (coordinates (1, 1.25)). We can observe diminishing returns, though, as higher speedups are very expensive to get. For instance, reaching e.g. 1.32x requires increasing the size of the RS by 25 % and those of other buffers by 75 %.

Interestingly, we can see that *reducing* the size of the RS can provide a speedup of 1.06x (coordinates (0.8, 1)). Similarly to the odd cases presented above for latency changes, this is due to how the dispatch order of uops can be changed in a coincidentally better way when degrading buffer sizes. However, such counter-intuitive cases are uncommon.

Furthermore, we can observe that decreasing the size of all buffers by 60 % (coordinates (0.4, 0.4)) causes a negative speedup of only 0.69x (i.e. a 31 % performance penalty).

We can hence easily determine the sweet spot for *performance per buffer entry* with UFS, as well as any degrees of compromise between small buffer sizes and best achievable ILP. However, other models and tools are needed to evaluate the consequences of such buffer size changes in terms of hardware complexity and power consumption.

13.6 Related Work

Code Quality Analyzer (CQA) [2], to which we compared UFS throughout this chapter, is the tool the closest to UFS that we know of: both analyze loops at a binary/assembly level, rely on purely static inputs and have a special emphasis on L1 performance. They actually both use the MAQAO framework to generate their inputs. CQA works in terms of bandwidth, which it assumes to be unimpeded by execution hazards. As its name suggests, it assesses the quality of targeted loops, for which it provides a detailed bottleneck decomposition as well as optimization suggestions and projections. UFS differs by focusing solely on time estimations, accounting for dispatch inefficiencies and limited buffer sizes. It does so by simulating the pipeline's behavior on a cycle-accurate basis, adding accuracy at the cost of speed. Finally, CQA supports more microarchitectures than UFS.

IACA [1] works similarly to CQA, and estimates the throughput of a target code based on uop port binding and latency in ideal conditions. It can target arbitrary code sections using delimiting markers, while both CQA and UFS only operate at the loop level. It does not account for the hazards UFS was tailored to detect, and we consequently expect it to be faster but less accurate. Like CQA, IACA also supports more microarchitectures than UFS.

Zesto [18, 19] is an x86 cycle-accurate simulator built on top of SimpleScalar [20] and implements a very detailed simulation of the out-of-order engine similar to that of UFS. However, as with other detailed simulators like [21], the approaches are very different: it works as a regular CPU simulator and handles the semantics of the simulated program. Its simulation scope is also much wider, with a detailed simulation of branch prediction, caches and RAM. UFS focuses solely on the execution pipeline, and particularly the out-of-order engine. It disregards the semantics, and targets loops directly with no need for contextual information (such as register values, memory state, etc.), making it considerably faster due to both not having to simulate regions of little interest and simulating significantly fewer things. Furthermore, UFS targets Sandy Bridge, Ivy Bridge and Haswell, while to the best of our knowledge Zesto only supports older microarchitectures.

Very fast simulators exist, but typically focus on different problematics. For instance, Sniper [22, 23] uses both interval simulation (an approach focusing on miss events) and parallelism to simulate multicore CPUs efficiently. As said events (cache misses and branch mispredictions) are irrelevant in the cases targeted by UFS (memory accesses always hit L1, loops have no if statements and have large numbers of iterations), the use cases are completely disjoint.

UFS is to our knowledge the only model targeting binary/assembly loops that both disregards the execution context and accounts for dispatch hazards and limited out-of-order resources.

13.7 Future Work

Evaluating the impact of unmodeled hardware constraints would be interesting to determine whether or not implementing them in UFS could be profitable. Such constraints include writeback bus conflicts and partial register stalls.

The impact of simulating fewer loop iterations should also be studied, as our current default value of 1000 may be unnecessarily high and time consuming.

As our base UFS model is aimed at Sandy Bridge, we could easily construct models for incremental improvements such as Ivy Bridge and Haswell on top of it. However, a validation work is necessary to evaluate their respective fidelities, and see if more microarchitecture-specific rules have to be implemented. Expanding the model to support further "Big Core" microarchitectures (e.g. Broadwell, Skylake...) would also be of interest.

The idea of Uop Flow Simulation can be applied to vastly different microarchitectures (such as the one used in Silvermont cores, or even ARM CPUs), and could have interesting applications beyond performance evaluation tools. For instance, its working out of context means it could easily be used by compilers to better evaluate and improve a generated code's quality.

In terms of codesign, UFS models could be used to quickly estimate the impact of a microarchitectural change on thousands of loops in a few minutes. Coupling this modeling technique with a bandwidth-centric fast-simulation model such as Cape [3] would allow for non-L1 cases to be handled efficiently as well.

13.8 Conclusion

We demonstrated UFS, a cycle-accurate loop performance model allowing for the static, out-of-context analysis of assembly loops. It takes into account many of the low-level details used by tools like CQA or IACA, and goes further by estimating the impact of out-of-order resource sizes and various pipeline hazards. It can also be used to evaluate how a loop would behave given different microarchitectural parameters (such as different out-of-order buffer sizes or load latencies).

Our Sandy Bridge UFS prototype shows that UFS is very accurate and exposes formerly unexplained performance drops in loops from industrial applications and in vitro codelets alike. Furthermore, it offers very high simulation speeds and can serially process dozens of loops per second, making it very cost effective.

Acknowledgments We would like to thank Gabriel Staffelbach (CERFACS) for having provided our laboratory with the AVBP application, as well as Ghislain Lartigue and Vincent Moureau (CORIA) for providing us with YALES2.

We would also like to thank Mathieu Tribalat (UVSQ) and Emmanuel Oseret (Exascale Computing Research) for performing and providing the in vivo measurements we used to validate UFS on the aforementioned applications.

This work has been carried out partly at Exascale Computing Research laboratory, thanks to the support of CEA, Intel, UVSQ, and by the PRiSM laboratory, thanks to the support of the French Ministry for Economy, Industry, and Employment through the COLOC project. Any opinions, findings, and conclusions or recommendations expressed in this material are those of the author(s) and do not necessarily reflect the views of the CEA, Intel, or UVSQ.

References

1. Intel: Intel architecture code analyzer (IACA) (2012). https://software.intel.com/en-us/articles/intel-architecture-code-analyzer
2. Oseret E et al (2014) CQA: a code quality analyzer tool at binary level. HiPC
3. Noudohouenou J et al (2013) Simsys: a performance simulation framework. RAPIDO'13, ACM
4. Palomares V (2015) Combining static and dynamic approaches to model loop performance in HPC. Ph.D. thesis, UVSQ, Chapter 7, Uop Flow Simulation
5. The AVBP code http://www.cerfacs.fr/4-26334-The-AVBP-code.php
6. YALES2 public page http://www.coria-cfd.fr/index.php/YALES2
7. MAQAO: Maqao project (2013) http://www.maqao.org
8. Fog A (2015) Instruction tables: lists of instruction latencies, throughputs and micro-operation breakdowns for intel, amd and via cpus. http://www.agner.org/optimize/instruction_tables.pdf
9. Djoudi L et al (2007) The design and architecture of maqao profile: an instrumentation maqao module. In: EPIC-6, IEEE, New York, p 13
10. Palomares V (2015) Combining static and dynamic approaches to model loop performance in HPC. Ph.D. thesis, UVSQ, Appendix A: Quantifying effective out-of-order resource sizes, Appendix B: Note on the load matrix
11. Intel: 2.2.2.4 (2014) Micro-op queue and the loop stream detector (LSD). Intel 64 and IA-32 Architectures Optimization Reference Manual
12. Intel: 2.2.4 (2014) The execution core. Intel 64 and IA-32 Architectures optimization reference manual
13. Intel: 3.5.2.4 (2014) Partial register stalls. Intel 64 and IA-32 Architectures optimization reference manual
14. Paoloni G (2010) How to benchmark code execution times on intel ia-32 and ia-64 instruction set architectures. Intel Corporation, Santa Clara
15. Koliaï S et al (2013) Quantifying performance bottleneck cost through differential analysis. In: Proceedings of the 27th international ACM conference on supercomputing. ACM, New York, pp. 263–272
16. Moureau V et al (2011) From large-eddy simulation to direct numerical simulation of a lean premixed swirl flame. Combust. Flame
17. Press WH et al (1992) Numerical recipes: the art of scientific computing
18. Loh GH et al (2009) Zesto: a cycle-level simulator for highly detailed microarchitecture exploration. In: IEEE International symposium on performance analysis of systems and software, 2009. ISPASS 2009, IEEE, New York, pp 53–64
19. Loh GH, Subramaniam S, Xie Y (2009) Zesto. http://zesto.cc.gatech.edu
20. Burger D, Austin TM (1997) The simplescalar tool set, version 2.0. ACM SIGARCH Comput Archit News **25**(3): 13–25

21. Binkert N et al (2011) The gem5 simulator. ACM SIGARCH Computer Architecture News **39**(2): 1–7
22. Carlson TE et al (2011) Sniper: exploring the level of abstraction for scalable and accurate parallel multi-core simulation. In: SC, ACM, New York, p 52
23. Heirman W et al (2012) Sniper: scalable and accurate parallel multi-core simulation. In: ACACES-2012, HiPEAC, pp 91–94

MIX
Papier aus verantwortungsvollen Quellen
Paper from responsible sources
FSC® C105338

If you have any concerns about our products,
you can contact us on
ProductSafety@springernature.com

In case Publisher is established outside the EU,
the EU authorized representative is:
Springer Nature Customer Service Center GmbH
Europaplatz 3, 69115 Heidelberg, Germany

Printed by Libri Plureos GmbH
in Hamburg, Germany